Simple M.S. Physical Science Investigations

Solutions for doing science in the classroom.

Christopher P. Garside

Seven Sides Publishing

Seven Sides Publishing has a mission to improve the teaching and understanding of science. To contact us, send an email to simpleinvestigations@sevensidespublishing.com or visit our website at sevensidespublishing.com.

Copyright © 2022 by Seven Sides Publishing and Christopher P. Garside. All rights reserved. No part of this publication may be reproduced, stored in a retrieval system, scanned, or transmitted in any form or by any means, electronic, mechanical, photocopying, recording, or otherwise, without the prior written permission of Seven Sides Publishing and Christopher P. Garside. Photocopying is permitted from this book to make copies only for the students of the teacher who owns this book; this is not for other teachers, students, or anyone else in the school or district's students.

ISBN: 9798420788509

Published by: Seven Sides Publishing, Cypress, TX.

Table of Contents

Introduction	**Page 4**
Unit 1 Properties of Matter *MS.PS1.A*	**Page 17**
Unit 2 Mixtures and Properties of Water *MS.PS1.A*	**Page 35**
Unit 3 Structure of Matter *MS.PS1.A*	**Page 70**
Unit 4 Chemical Reactions *MS.PS1.B*	**Page 91**
Unit 5 Acids and Bases *MS.PS1.B*	**Page 111**
Unit 6 Motion *MS.PS2.AB*	**Page 121**
Unit 7 Forces *MS.PS2.AB*	**Page 141**
Unit 8 Mechanical Energy *MS.PS3.ABC*	**Page 174**
Unit 9 Thermal Energy *MS.PS3.ABC*	**Page 204**
Unit 10 Electromagnetism *MS.PS2.B 3ABC*	**Page 223**
Unit 11 Waves *MS.PS4.ABC*	**Page 257**
NGSS Correlations	**Page 300**
Equipment List for all Investigations	**Page 306**

Introduction

To help teachers teach science through investigations, Seven Sides Publishing has provided a series of lab manuals for Elementary Science, Middle School Science, Biology, Chemistry, Physics, Earth & Space Science, and Environmental Systems. These manuals are a rich resource for structure and investigations. There is a shortage of user-friendly labs that easily allow teachers and students to perform experiments quickly. Too many labs have too much busy writing within them, where teachers do not want to take the time to read everything to figure out if it would be good for them to use with their students. If the teachers do not want to read it, do you think the students do? So we have taken a lot of the traditional labs that have been around for decades and simplified them; so they are easy to read and perform. We have also added some new original labs that have never been seen before. There have been efforts to have teachers do more investigations with their students, but there is no plan or solution to deal with the real issues teachers have in preparing to do this. The book How to Teach Science Through Investigations has the plan, and the Simple Investigations Lab manuals have the solutions so students can learn science through investigations with minimal effort. Teaching science through investigations will make your classrooms more efficient, where students learn content and practice skills simultaneously. Science is a process of doing. Doing this process is the most efficient way for students to learn science and be able to use it in the future. We live in a culture where science-literate people are needed for jobs, but too few can be found. If you incorporate these labs with virtual labs (that I will point you to in each section of the lab manual), skill/math practice, and concept maps, you will not need to fill in gaps by giving lectures. All content can be learned through investigations and practice. Remember, we only remember 5-20% of what we hear. That 20% is when you are really interested in the content. But hearing practices no science process skills and does not activate any higher cognitive thought. Lecturing is not a good option. We remember 75-80% of what we do/experience and 90-95% of what we teach. Investigations allow us to keep our students in these higher retention percentages. Teaching through investigations also works because students spend more time in class at higher Bloom's Taxonomy levels, staying in zones C and D on the Rigor Relevance Chart when they perform investigations. And if you add the physical way they are stimulated with the hands-on experience, you cannot deny the level of learning will be much higher while students perform investigations. This manual gives you the resources you need to teach Middles School Physical Science through investigations.

We separated each of these sections in the manual like you may divide your class units. We will follow the Next Generation Science Standards to make it easy for you to find the labs you want and need for your classes. We include concept maps at the front of each section that shows the vocabulary and visual clues to how concepts relate to each other; this is a great way to organize information. It talks to the students to see how ideas work together, making it easy to chunk information to use at higher cognitive levels. At the beginning of each lab, we put the materials you will need in boldface in the directions; this saves time for your lab preparation. There is also a safety question in boldface just after that for you and your students to evaluate. It says, "Looking at the material and lab we will be using, what are the safety precautions we should take to protect ourselves and materials during this investigation." Make sure to read the lab so you can better answer this question with your students.

Virtual Labs

Hands-on labs are not the only way for students to learn science, but they are the most effective. However, virtual labs can be used with these hands-on labs. Many investigations physically cannot be done hands-on, so some experiments will have to be done virtually. There are three sources that I have used in the past that have a good number of resources. **Physicsclassroom.com** and **PhET.colorado.edu** are free to everyone and are great to use. **Physicsclassroom.com** has teacher notes and activities/exercises that guide students through Physics and Chemistry Interactives. You can find them under the simulation and open, download, or print the PDF. They also have a series of Concept Builders that are a tremendous virtual practice that can replace those worksheets that help students practice concepts, math, and skills. They can be hard to find, so above the list provided is the section where they can be found (underlined and in italics) on the website. **PhET.colorado.edu** has a variety of activities of different levels that you can explore to go through their simulations. They are also easy to download and print. **ExploreLearning.com** is expensive, but the quality of its product is much higher than the other two. When you click on a Gizmo, you can also click on lessons and find the Student Explorations that go with each Gizmo that you can modify, download, and print. They are written at a very high quality, making the students think like a scientist. At the end of each section of this lab manual, we include a list of virtual labs from these organizations that would be great to use with these labs. Please remember virtual labs should never replace hands-on labs. If the students can learn the content live, that should be the priority because it is more of an experience that will be remembered. There are many other virtual simulations out there, but none so far have moved me to use them over the three I have mentioned here.

TIPPERs

TIPPERs are great for students to explore and think about different scenarios for each concept of Physics and Chemistry. These help students think outside the box, apply concepts to real life, and think about how multiple concepts would be used together. I suggest you get the books of TIPPERs to practice and discuss after completing these labs and investigations.

Probe-ware

This lab manual has lots of labs that use probe-ware. Students must learn how to use probe-ware; this means teachers need to know how to use probe-ware. Many companies use digital probe-ware with all the research, development, testing, and forensic testing they do; this has potential career opportunities that help students become more marketable for jobs if they are familiar with using probe-ware. Hooking everything up is just as easy as charging your phone. When I was a High School Science Technology Coach and researched which companies and devices would be the most user-friendly to students, I found using Vernier Probe-ware was better for high school students, but PASCO seemed better for middle school students. Both are giants in the probe-ware industry for education. Since I am more familiar with Vernier, I will be referring to Vernier Probe-ware. However, PASCO would be a great alternative.

Interfaces are devices that the probes are connected to that talk with the program (Logger Pro) that displays the data. I found the most economical and friendliest way for students to see the data from probe-ware is to use the Vernier LabQuest Mini interface hooked up to a computer with Logger Pro. LabQuest Mini has multiple ports that are needed in many labs. They are the least expensive, so they are better on the budget. They require no batteries, so they are easy to transport if you need or want to. The other interfaces are more expensive, require batteries if you are going outside, or the stand-alone devices have a smaller screen to see the data, with less flexibility to manipulate the parameters like changing the time of data collection or changing units if you want to change or modify an experiment. Some costly wireless probes and interfaces may be easier to use if you do not mind the cost. A computer screen is much bigger and makes it easier to see the data, so this is my preferred setup. But using any interfaces will work fine for these labs.

Connecting the Probe-ware

To hook them up, you will plug your probe into one of the channels or the sonic on the interface. If the plug does not fit in smoothly, either you are plugging it upside-down or trying the wrong port. Then take the little chord that looks like it would go into your phone and plug that into your interface. Take the other end, and plug it into a USB port on your computer. Open up Logger Pro on your computer. If everything is hooked up properly and the computer and interface are working properly, you will see a green button at the top of the computer screen that says "Collect." Many of the labs have preset settings in Logger Pro. You will use the manila folder at the top left of the toolbar in Logger Pro to find the folders and files you will be instructed to go to for these specific settings for different labs. Whenever you get the physical equipment, they will have detailed instructions in the box they come in on how to hook them up if you are still confused. They will also have instructions on how to calibrate the probes if needed. There are a few probes that require frequent calibration. If we use any, it will be discussed in the lab directions. The more you use probe-ware, the easier it gets to set up. I usually only have to show my students twice to have them be able to set the equipment up on their own. But as you are showing them, have them physically do it. You can also find detailed instructions online at Vernier.com. Many more detailed labs can also be found there under lab ideas.

You also can use standard equipment like spring scales for force sensors or thermometers for temperature probes. Because schools want to integrate more technology, we wrote these labs to use probe-ware where ever they were applicable. Because they are so simple, these labs can be modified to fit whatever equipment you have. There are very few labs that I have used in my career that I did not alter how I presented them. One reason we wrote these labs this way is to customize them to the Texas TEKS and National Standards. We also wrote them how we thought a teacher would want to use them.

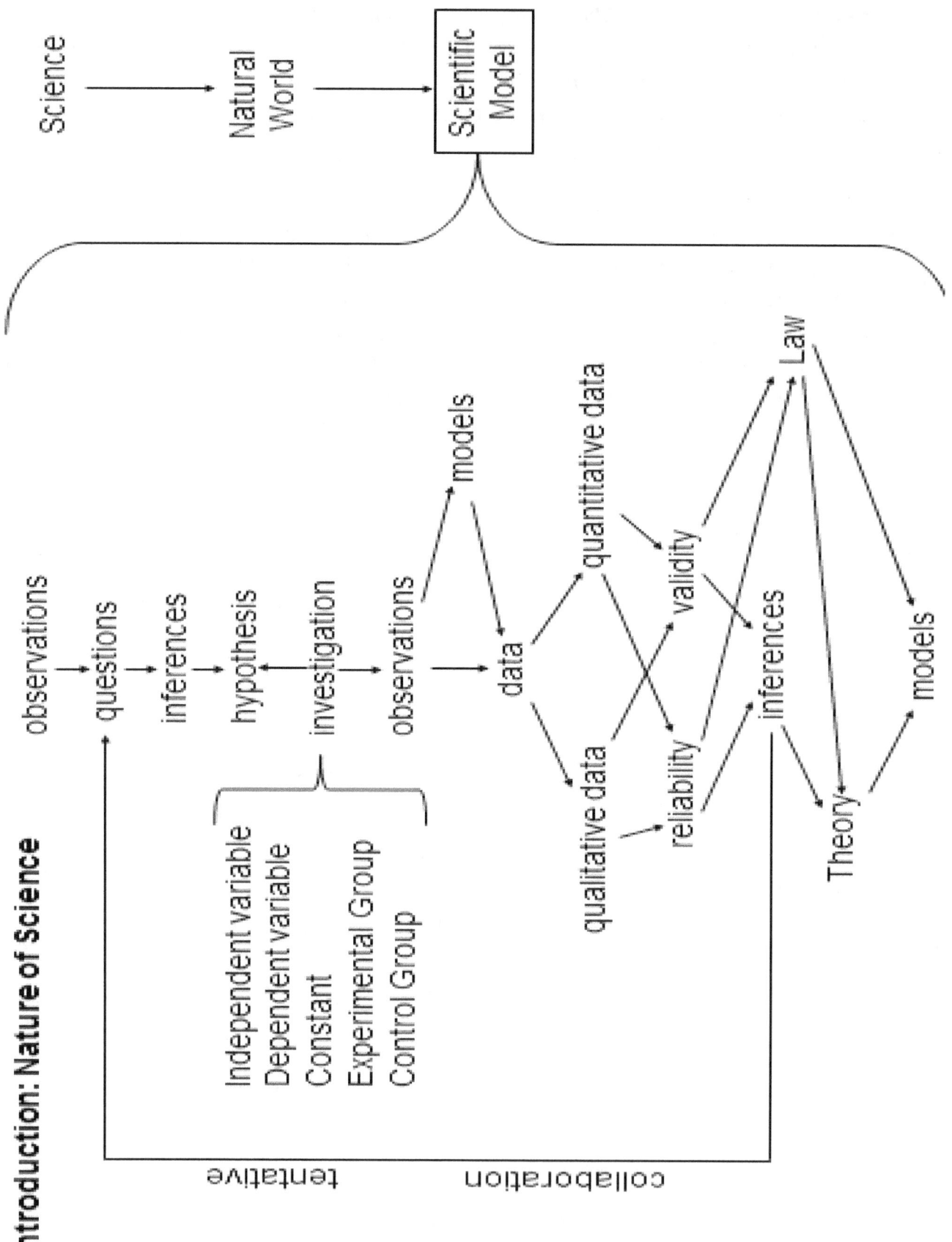

Focus on the Process

Directions:

Get a **small Legos set**. Teachers, make sure it is not too easy for your students. You are going to try to put it together in two different ways. Time how long it takes to put it together each way and answer the questions that follow. **Looking at the materials and lab we will be using, what are the safety precautions we should take to protect ourselves and materials during the investigation?**

   A) Take the Lego pieces and construct the picture (the **product**) on the box's cover, looking at nothing but the cover and the Lego pieces.

   B) When 20 minutes have passed, or you are done, take what you have made totally apart. Take out the directions (the **process**) and construct the product while using the step-by-step directions. Time how long it took you to complete the set.

Questions:

1) How did it feel trying to construct the Legos (A) without any directions?

2) Did you finish? If so, how long did it take?

3) How did it feel to construct the Legos (B) with step-by-step directions?

4) Did you finish? If so, how long did it take?

5) Which strategy (A or B) allowed you to complete the product?

6) Which strategy (A or B) was more intimidating?

7) Which strategy (A or B) allowed you to see what is under the surface?

8) Which strategy (A or B) will allow you to learn more?

We often get anxious or procrastinate when faced with a large task. We are tempted to take a "shortcut" (copy or cheat, we do not learn much when we do this). There are pain and stress hormones that are released when this happens. One way to overcome this is to just worry about the next step in the process and not worry about the product. You can see and measure progress, which makes the process not feel too bad. Another way is just to start working. When you start working, those pain and stress hormones stop getting released so that anxiety goes away; this is why when we want to learn efficiently and effectively, we must:

Focus on the _____ and the _____ will take care of itself.

9) How is putting the Lego pieces together like putting ideas together to understand concepts?

Simple Middle School Physical Science Investigations   Seven Sides Publishing

Measurement Lab

Directions:

You will need **water**, a **scale**, a **meter stick**, a **temperature probe** attached to an **interface** connected to a **computer** with **Logger Pro**, a **100 mL graduated cylinder**, and a **stopwatch**. **Looking at the materials and lab we will be using, what are the safety precautions we should take to protect ourselves and materials during the investigation?**

1) Take the graduated cylinder and find its mass empty; write this in Data Table 1.
2) Add 50 mL of water to the graduated cylinder. Make sure you use the meniscus properly where the volume is at the bottom of the meniscus. Have the teacher check that you measured it correctly. Have each person in your group empty and fill the graduated cylinder with 50 mL of water. As they do so, have each person in your group time how long it takes for each person to fill the graduated cylinder and check it is correct (it is not a race, just a chance to get familiar with using the graduated cylinder and stopwatch).
3) Now find the mass of the graduated cylinder with 50 mL of water in it. Subtract the mass of the empty graduated cylinder from this mass and write the water's mass in Data Table 1.
4) Connect your temperature probe to an interface and connect your interface to a computer with Logger Pro (unless you have a LabQuest 2, then just hook your probe to the LabQuest 2). Find where the Logger Pro is located on your computer so you can use it again in the future. Once open, find the graduated cylinder's water temperature in Fahrenheit and Celsius (you will have to figure out how to change units). Write these in Data Table 1.
5) Take your meter stick and measure the length of the graduated cylinder. And measure the width of the base in centimeters. Write these in Data Table 1

Data Table 1

Object	Mass (g)	Volume (mL)	Time to Fill (s)	Temp (°F)	Temp (°C)	Length (cm)	Width (cm)
Graduated Cylinder		✖		✖	✖		
Water			✖			✖	✖

Simple Middle School Physical Science Investigations   Seven Sides Publishing

Questions:

1) Convert a length to meters, the volume to liters and a mass to kilograms, and Celsius to Kelvin.

   Length _____ m   Volume _____ L   Mass _____ kg   Temp _____ K

2) What do you notice about the mass of the water compared to its volume?

3) What can happen to your investigations if your measurements are not accurate or precise?

4) Why do you think the rest of the world uses the metric system over the English system.

Patterns in Pennies

Directions:

You will need a ruler, 10 pennies, a scale, a roll of pennies, and an empty penny roll. Looking at the materials and lab we will be using, what are the safety precautions we should take to protect ourselves and materials during the investigation?

1) Find the mass of one penny with a scale to the nearest .1 g. Then measure the height of the penny in millimeters. Write these in Data Table 1 below.
2) Place another penny on top of the original penny and find the mass and height of the two pennies. Write these in Data Table 1 below.
3) Keep adding pennies one by one, measuring the mass and height until you have 10 pennies on the scale.
4) Make a line graph with the mass on the (x) axis and the height on the (y) axis for the pennies on Graph 1.

Data Table 1

Number of Pennies	Mass	Height
1		
2		
3		
4		
5		
6		
7		
8		
9		
10		

Graph 1

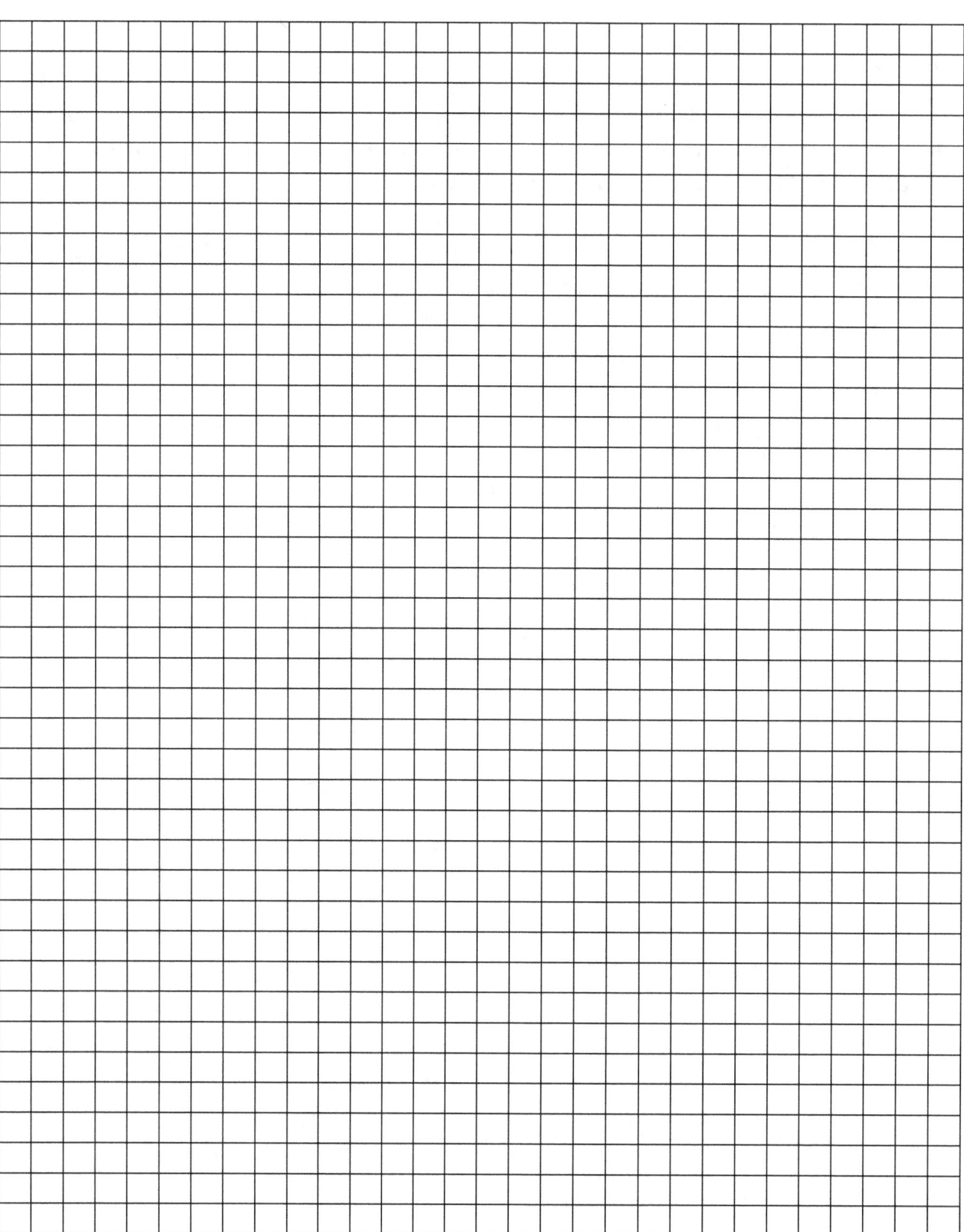

Questions:

1) What do you notice about the graph?

2) Is this a direct or inverse relationship between mass and height?

3) Do all pennies have the same mass? (Explain)

4) Do all the pennies have the same thickness? (Explain)

5) Use your data to estimate how many pennies are in the coin roll. How many pennies do you think are in the roll?

6) What did you do to estimate the number of coins?

7) What else could you do to estimate the coins?

8) Try your answer to #7. Do you get the same number as #5?

9) Carefully open up the coin roll and find out how many pennies there are. How close were you to the real number? After you are done counting, carefully close the roll back up.

10) Calculate the % accuracy by taking the lowest number between your guess and the actual number dividing by the higher of the two, then multiplying by 100.

11) What were some sources of error?

Virtual Investigations to go with Introduction

ExploreLearning.com

- Unit Conversions Gizmo
- Graphing Skills Gizmo
- Measuring Volume Gizmo
- Elevator Operator (Line Graphs) Gizmo
- Weight and Mass Gizmo
- Triple Beam Balance Gizmo
- Reaction Time 1 Gizmo
- Reaction Time 2 Gizmo

Physicsclassroom.com/Concept-Builders/Chemistry:

- Measurement and Numbers
- Significant Digits and Measurements
- Metric System
- Metric Estimation
- Experiments and Variables
- Proportional Reasoning
- Calculating Slope
- Using Graphs
- Which One Doesn't Belong

Unit 1: Properties of Matter

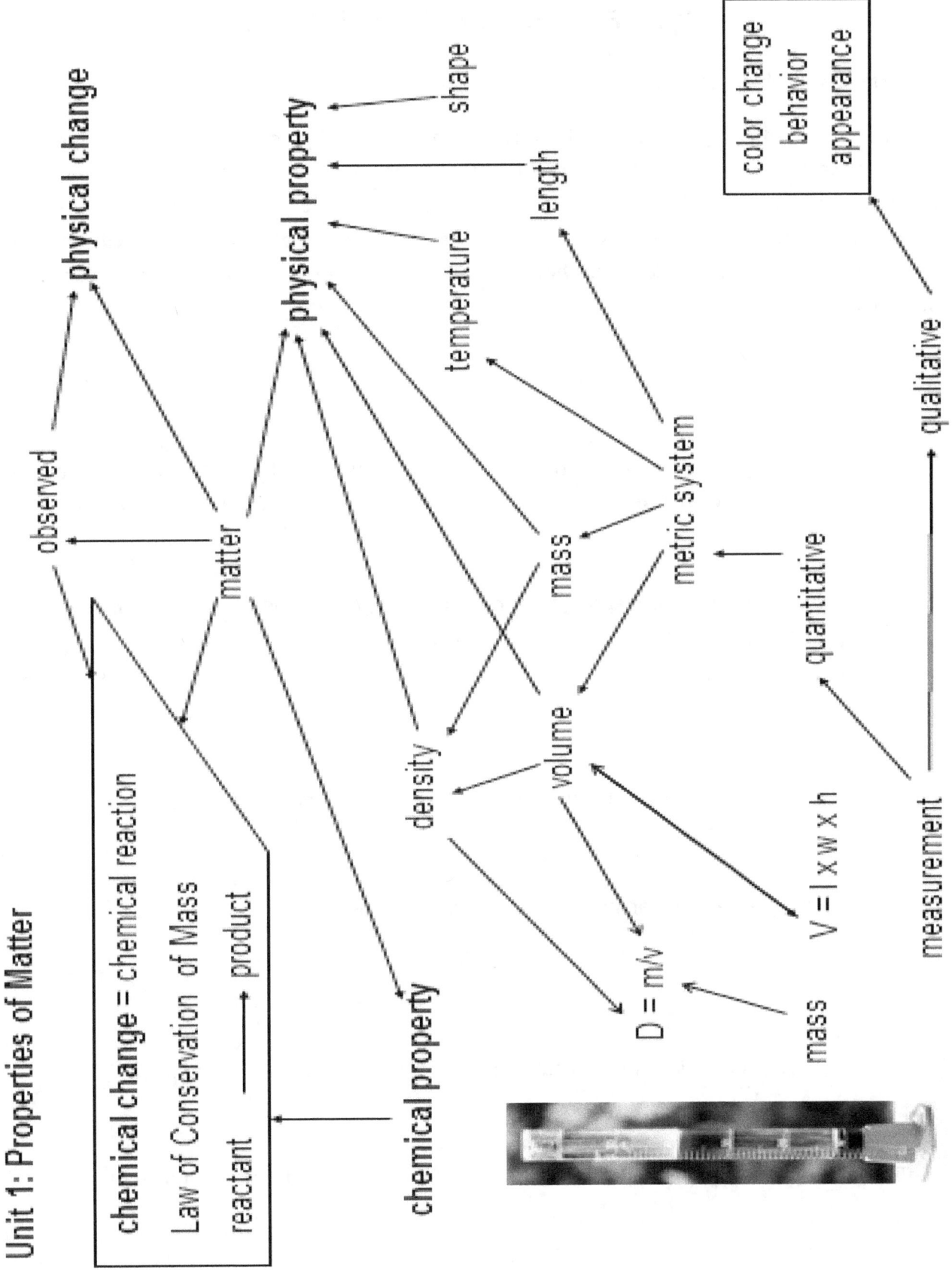

Simple Middle School Physical Science Investigations   Seven Sides Publishing

Physical and Chemical Changes

Directions and Observations:

You will need **safety goggles**, an **aluminum pan**, **paper**, a **pipette**, a **candle**, a **lighter** or **matches**, **salt**, a **beaker of water**, **vinegar**, **baking soda** in a **small beaker**, **steel wool**, **long forceps**, a **clean piece of metal** and **another that is corroded** (of the same kind of metal), **hydrogen peroxide**, and **liver** or **banana**. Looking at the materials and lab we will be using, what are the safety precautions we should take to protect ourselves and materials during the investigation?

1) **Physical changes** are changes that happen to a substance that does not cause any new substances to form. **Chemical changes** are changes that occur when new substances are formed.

2) Make sure everyone is wearing their protective goggles. Have your teacher come by and light the candle. What do you see happening to the wick of the candle?

3) Is this creating any new substances? Fill in Data Table 1 for the burning wick.

4) What do you see happening to the wax of the candle?

   a. Is this creating any new substances? Fill in Data Table 1 for the melting wax.

5) Take a piece of paper and tear it up. What did you see happen to the paper?

6) Did tearing the paper create any new substances? Fill in Data Table 1 for tearing paper.

7) Wad up a piece of paper and put it in the aluminum pan. Make sure everyone is wearing their protective goggles. Have your teacher come by with the lighter and light the paper on fire. Make sure to keep everything away from the flames. What do you see happening to the paper?

   a. Is it creating any new substances? Fill in Data Table 1 for burning paper.

8) Pour the salt into a beaker of water and stir it up. What do you see?

   a. Did this create any new substances? Fill in Data Table 1 for dissolving salt.

9) Now pour some vinegar into the small beaker of baking soda. What happened?

   a. Did this make any new substances? Fill in Data Table 1 for mixing vinegar and baking soda.

10) Look at the two pieces of metal. What difference do you see between the clean metal and the corroded?

11) Did corrosion cause a change to create a new substance? Fill in Data Table 1 for the metal.

12) Make sure you are wearing your safety goggles. Look at your steel wool and hold it with your forceps. Now have your teacher expose it to fire with the lighter. What do you see happen? Look at the color when your teacher is done (only the part of it that was lit).

   a. Did this produce any new substances? Fill in Data Table 1 for burning steel wool.

13) Finally, take some hydrogen peroxide and pour it on some liver or a slice of banana. What do you see happen?

   a. Did this produce a new substance? Fill in Data Table 1 below for hydrogen peroxide on life.

Data Table 1

Change that Happened	Physical or Chemical Change?	Evidence Why
Burning wick		
Melting wax		
Tearing Paper		
Burning paper		
Dissolving salt		
Vinegar and baking soda		
Corrosion of metal		
Burning of steel wool		
Hydrogen peroxide on life		

Questions:

1) Describe observations you might see when a physical change occurs.

2) Describe observations you might see when a chemical change occurs.

3) How could you tell dissolving sugar in water is a physical change?

4) Why does bubbling let you know there is a chemical change?

5) How do you know burning something is a chemical change?

6) Are melting and boiling physical or chemical changes?

Student Atomic Motion

Directions and Questions:

You will need a class of **students** (that will act like atoms as a solid, liquid, and gas). **Looking at the materials and lab we will be using, what are the safety precautions we should take to protect ourselves and materials during the investigation?**

1) Have students sit in their seats. They are now modeling how atoms move when they are in a **solid-state**. Are they absent of any motion? If not, describe how the student molecules are moving.

2) Now have the students get up and walk slowly around in a small area of the class; they are now modeling **liquid** atoms. How is this motion of the **liquid state** different from the solid state?

3) Now have the students walk around the classroom faster over the entire classroom. When they are about to bump into another person, have them not touch; they just move quickly in another direction. The students are now modeling how atoms move in a **gaseous state**. How do the **gas** atoms move differently from the other two states?

4) Which state of matter had the most energy moving in it? Explain how you could tell.

5) Which state of matter had the least energy moving in it? Explain how you could tell.

6) How could you measure the density of the molecules in this activity?

7) Which state of matter had the smallest density?

8) Which state of matter had the largest density?

9) Which states of matter could change shape to fill the container?

10) Which state of matter could not change shape?

11) Which state of matter was able to fill the whole container?

12) How could we change this model to show molecules (atoms bonded together) in motion as solid, liquid, and gas?

13) How was this model not accurate in showing atomic motion?

Observing Molecular Motion

Directions:

You will need at least three of the **biggest beakers** in your school. They all will be filled with water. One beaker **filled with water** needs to be put in a **refrigerator** so the water is cold. You will also need one beaker on a **hotplate** before doing the demo, so it heats up (if it boils, you can turn the heat off). The third beaker will be room temperature water straight out of the tap. Lastly, you will need some **food coloring. Looking at the materials and lab we will be using, what are the safety precautions we should take to protect ourselves and materials during the investigation?**

1) Line all three beakers up from coldest to warmest where the whole class can see. Place a drop of food coloring in the cold beaker and have the students watch how the food coloring spreads.
2) Put a drop of food coloring in the room temperature water, then in hot water on the hotplate. Have students observe the movement in all three beakers. It will not take long for the hot one to become homogeneous.
3) Have the students draw what they see in the three beakers below.

        Cold   Warm   Hot

Questions:

1) In which beaker did the dye move the fastest?

2) In which beaker did the dye move the slowest?

3) Why do you think this happened this way?

4) Were the substances in the beakers a solid, liquid, or gas? Explain how you know.

5) When was the substance in the beakers pure?

6) When was it a heterogeneous mixture?

7) When was it a homogeneous mixture?

8) Temperature is defined as the average kinetic energy of molecules. How does this explain what was happening in the hot, warm, and cold water?

9) Which beaker had the most energy?

10) How can this explain why we get hurt when we touch something hot?

Seeing the Heating Curve (Physical Change)

Directions:

You will need a **beaker**, **frozen water**, and a **temperature probe** suspended in it; this needs to be prepared for the day before in the **freezer**. Freezing water in the beaker many times breaks the beakers, so freeze the water in paper or **Styrofoam cups** while suspending the temperature probes in the water. When doing the investigation, peel off the cup and put the ice in the beaker. Attach the probe to an **interface** that is connected to a **computer** with **Logger Pro**. You will need to place the beaker of ice on a **hotplate**. You also need a **ring stand** and **clamp** to hold the temperature probe up as the ice melts. **Looking at the materials and lab we will be using, what are the safety precautions we should take to protect ourselves and materials during the investigation?**

1) Make sure the Logger Pro is set up to collect temperature data every second for at least 20 minutes.
2) Click "Collect" and turn the hotplate on high.
3) Watch the setup and data until the ice melts to water and the water boils for a little while. Then click "Stop."
4) Use the graph the Logger Pro gives you to answer the questions below.

Questions:

1) What was the temperature range of the ice in the data?

2) What did the graph look like while the water was melting?

3) Was this a physical change or a chemical change?

4) What was the temperature range while the water was liquid?

5) As the water boiled, what was the temperature?

6) What does the graph look like during this phase change?

7) What do you think the temperature range would be for steam?

8) Build and label Graph 1 of what the graph looked like on the Logger Pro.

Graph 1

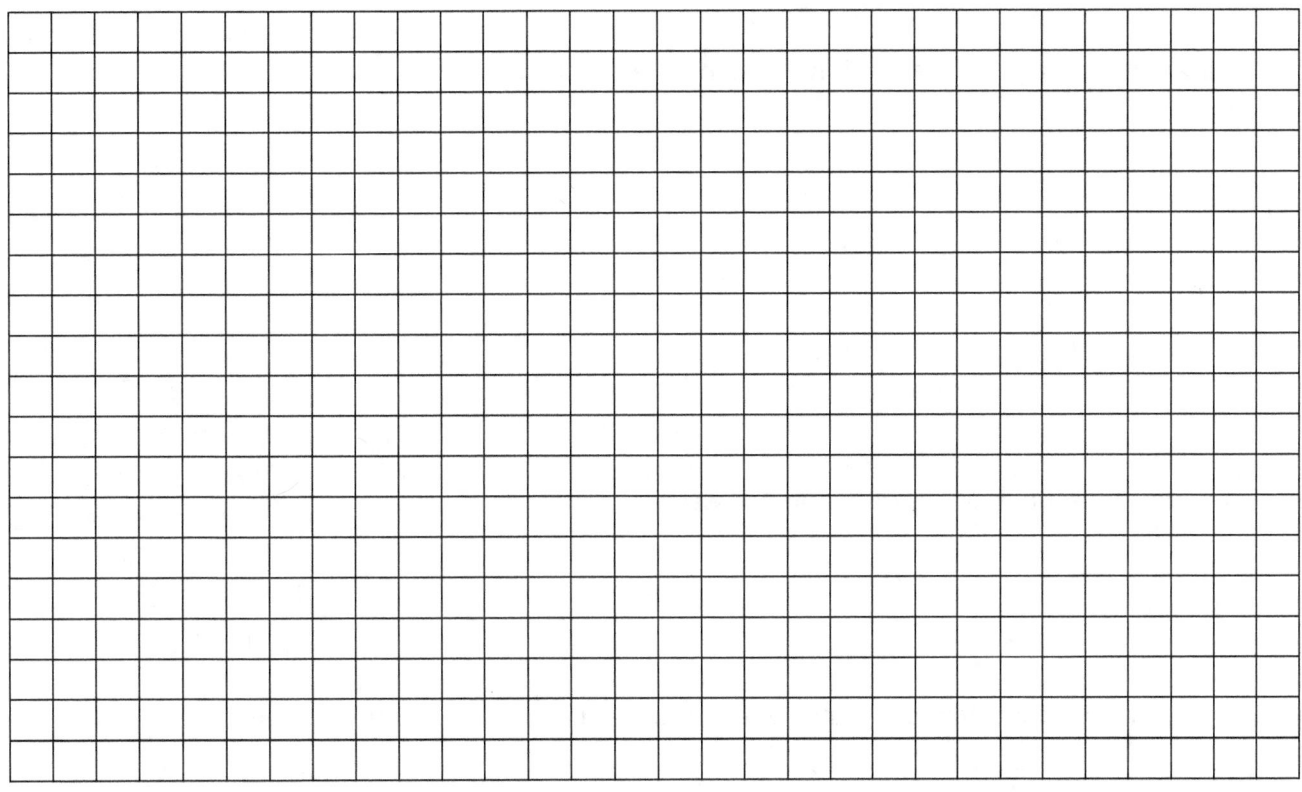

Convection in Liquids and Gases

Directions:

Fill a **large beaker** with **water**, add **pepper** to it, put on your **safety goggles**, place the beaker on a **hotplate**, and heat the water to just below the boiling point. **Looking at the materials and lab we will be using, what are the safety precautions we should take to protect ourselves and materials during the investigation?**

1) Draw a picture of the motion of the pepper in hot water:

2) Describe the motion you see in hot water:

3) How does the change in water density cause this motion (go into detail)?

4) Light a **candle** with a **match/lighter** and gently blow it out. Which direction does the smoke go?

Questions:

1) Describe why the particles of the pepper moved the way they did as the water became hotter.

2) Describe why the pepper particles moved the way they did as the water became colder after losing heat on the surface.

3) Explain how convection currents formed in the beaker.

4) Explain why the motion of the particles changed as the burner heated up the water.

5) Were the contents in the beaker a pure substance, a homogeneous mixture, or a heterogeneous mixture? Explain why.

6) Were the contents of the beaker a solid, liquid, or gas? Explain why.

7) Why did the smoke go that direction when the candle was blown out?

Which is Denser in the Mixture?

Directions and Questions:

Have a **bottle** filled with **oil** and **water. Looking at the materials and lab we will be using, what are the safety precautions we should take to protect ourselves and materials during the investigation?**

1) The liquid with the biggest density would be the liquid on the bottom, and the liquid with the smallest density would be on the top. Which liquid has the biggest density?

2) Which liquid has the smallest density?

3) Water is made up of all Hydrogen and Oxygen. Oil is made of mostly Hydrogen and Carbon and a small amount of Oxygen. Use a periodic table looking at the masses of these elements and where they sit in the periodic table, then explain why one floated on top of the other.

4) Is this a heterogeneous or homogeneous mixture? (Hint: the prefix: **hetero** means different, and the prefix: **homo** means same)

5) What are the pure substances in the bottle?

The Density of Oddly Shaped Objects

Directions:

You will need a **scale**, **water** in a **graduated cylinder**, and **any objects** that will sink in water and fit in your graduated cylinder. **Looking at the materials and lab we will be using, what are the safety precautions we should take to protect ourselves and materials during the investigation?**

1) Use a scale to find the mass of your first object. Write the name and the mass of this object in Data Table 1.
2) Fill your graduated cylinder halfway with water. Write this measurement in Data Table 1.
3) Carefully place your object in the graduated cylinder without splashing any water out. The easiest way to do this is to tip the graduated cylinder (without letting any water out) and gently slide the object into the water using the graduated cylinder's inside surface.
4) Measure the volume of the water now and place that in Data Table 1.
5) Subtract the initial volume from the final volume; this will be the object's volume. Write this in Data Table 1.
6) Now calculate the object's density by taking the mass and dividing it by the volume. Write this density on the right column of Data Table 1.
7) Repeat steps 1-6 for all your objects and write the data in Data Table 1.

Data Table 1

Object	Mass (g)	Initial Volume (mL)	Final Volume (mL)	Volume of Object (mL)	Density of Object D=m/V
					g/mL
					g/mL
					g/mL
					g/mL
					g/mL

Questions:

1) Which object is the densest?

2) Which object is the least dense?

3) Why did the volume of water appear to rise when you put in your objects?

4) What is another way of measuring the volume of objects that are not irregularly shaped?

5) What if the object you want to measure floats in the water; how can you find its density?

6) What do you think affects the density of objects?

Virtual Investigations that go with Properties of Matter

ExploreLearning.com:

    Phase Changes

    Melting Points

    Phases of Water

    Temperature and Particle Motion

    Diffusion

    Polarity and Intermolecular Forces

    Chemical and Physical Changes STEM Case

    Chemical and Physical Changes Handbook

    Properties of Matter STEM Case

    Properties of Matter Handbook

    Density Laboratory Gizmo

    Density via Comparison Gizmo

    Density Gizmo

    Density Experiment: Slice and Dice Gizmo

    Determining Density via Water Displacement Gizmo

    Archimedes' Principle Gizmo

Phet.colorado.edu:

    Density

    Concentration

    Salts and Solubility

    Sugar and Salt Solutions

    Energy Forms and Changes

States of Matter

States of Matter Basics

Physicsclassroom.com/Concept-Builders/Chemistry:

Classification of Matter

Chemical vs. Physical Properties

Density Ranking Tasks

Unit 2: Mixtures

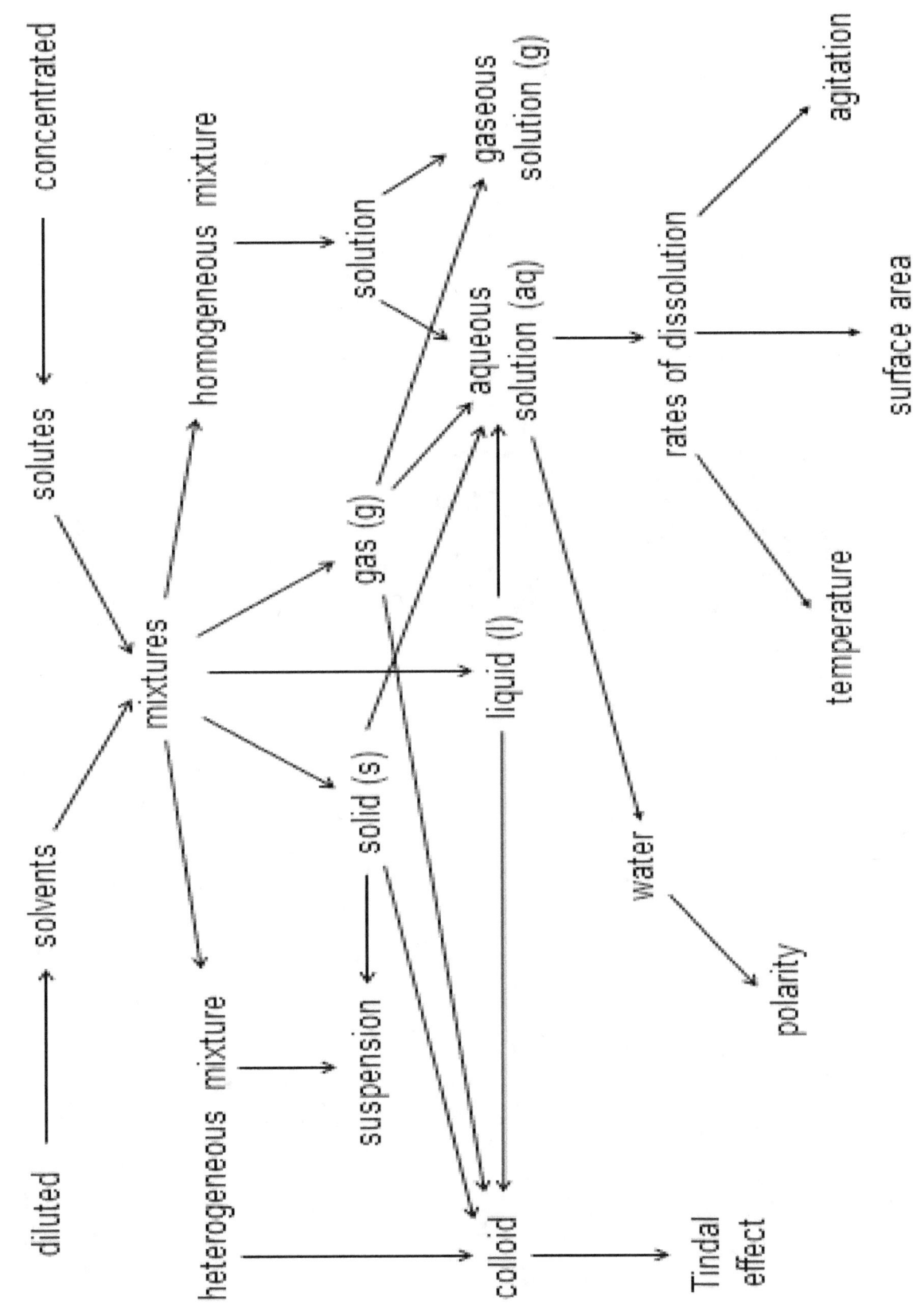

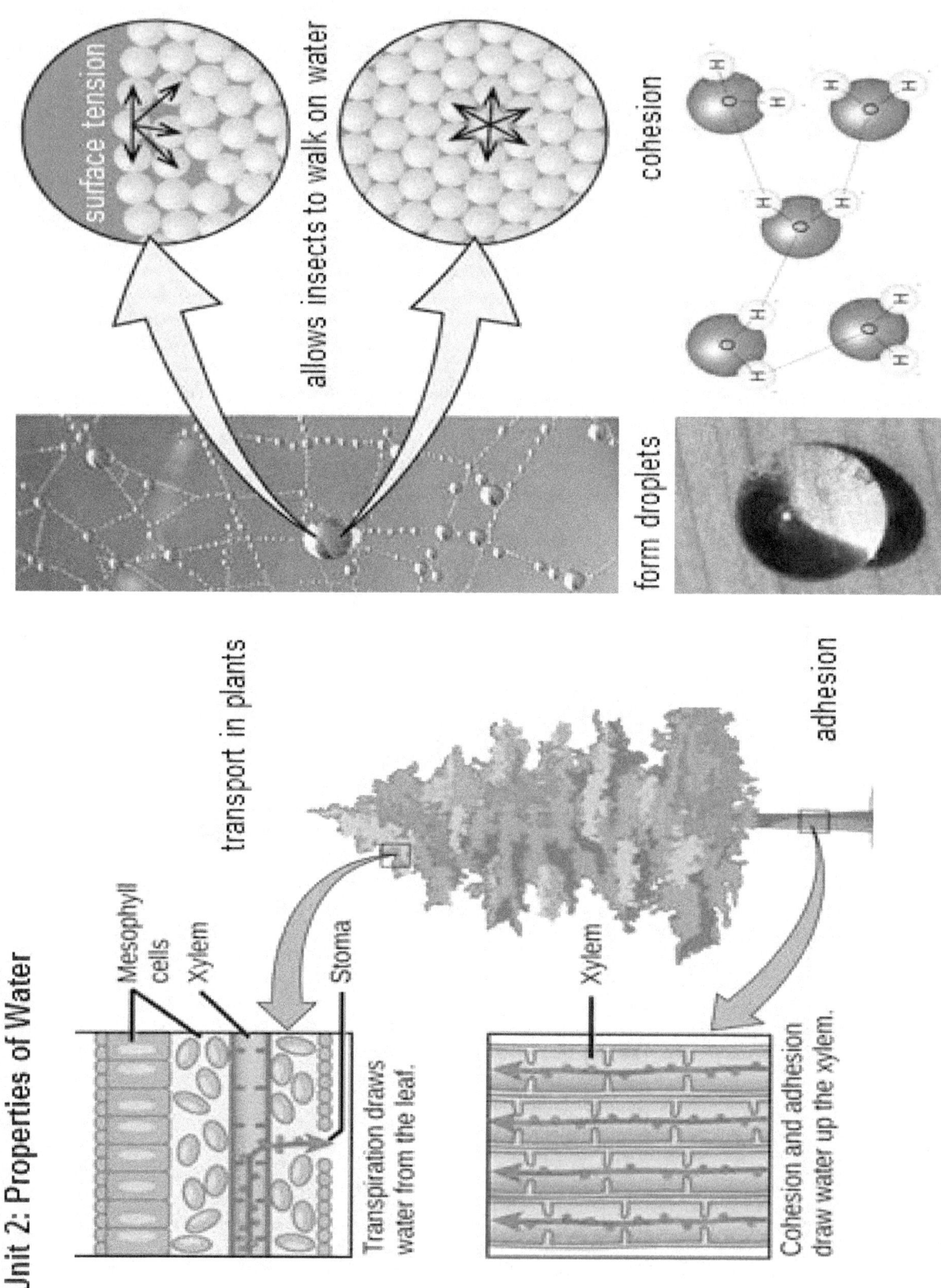

Elements Compounds and Mixtures Research

Directions and Questions:

Using your teacher's instructions, use the **internet** and your **textbook** to research elements, compounds, and mixtures, then answer the following questions.

1) What is an element, and how is it related to compounds and mixtures?

2) What are examples of elements?

3) What is a compound, and how is it related to elements and mixtures?

4) What are examples of compounds?

5) What is a mixture, and how is it related to elements and compounds?

6) What are the different categories of mixtures, and how do you tell them apart?

7) How were the different elements made?

8) Can an element be separated?

   a. Is it easy or hard? Explain.

9) How can a compound be separated?

   a. Is it easy or hard? Explain.

10) How can a mixture be separated?

    a. Is it easy or hard? Explain.

11) What does this investigation tell us about ourselves?

12) Fill in the Ven diagram below comparing and contrasting elements, compounds, and mixtures.

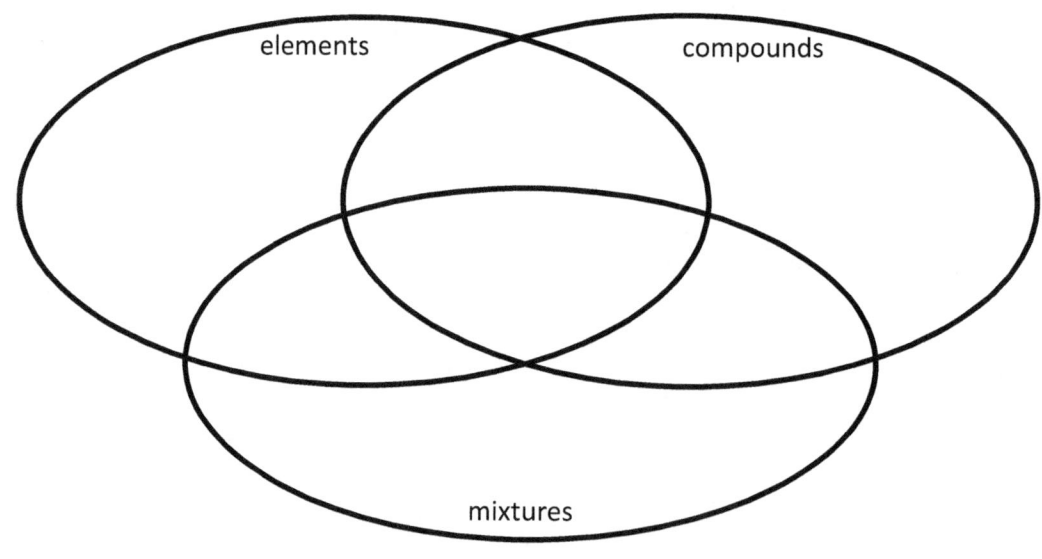

Simple Middle School Physical Science Investigations   Seven Sides Publishing

Elements Compounds and Mixtures

Directions:

You will need **aluminum foil**, **milk** in a **small beaker**, a **laser pointer**, **water** in a **small beaker**, **granite countertop samples**, **Kool-Aid** in a **small beaker**, **salt** (sodium chloride), a **pencil**, **chalk** (calcium carbonate), and **muddy water** in a **small beaker**. Looking at the materials and lab we will be using, what are the safety precautions we should take to protect ourselves and materials during the investigation?

1) **Pure substances** can be **elements** or **compounds**.
    a. **Elements** are pure substances with only one type of atom found on the periodic table.
    b. **Compounds** are pure substances with two or more elements in a fixed ratio.
2) **Mixtures** have varying ratios of different substances.
    a. **Homogeneous mixtures** appear the same throughout. They are also called **solutions**. Lasers can go through **solutions**.
    b. **Heterogeneous mixtures** are different throughout. Lasers are deflected in heterogeneous mixtures.
        i. **Colloids** look like homogeneous mixtures but have large particles that do not settle out.
        ii. **Suspensions** have larger particles that do settle out.
3) Analyze each substance in front of you and classify it as either a pure substance or a mixture. Then classify the type of pure substance or mixture. This data will be filled in Data Table 1.

Data Table 1

Object	Identity (Pure or Mix)	Classification
Aluminum foil		
Milk		
Water		
Salt		
Granite countertop		
Kool-aid		
Salt		
Pencil		
Chalk		
Muddy water		

Simple Middle School Physical Science Investigations   Seven Sides Publishing

Questions:

1) If you know the name of a substance, how can you tell whether it is an element or not?

2) How did you find out how milk was classified?

3) How would you classify a human? Explain why.

4) How would you classify fog?

5) How would you classify a clean atmosphere?

6) How would you classify smog?

7) How would you classify brass?

Separating Mixtures

Directions and Questions:

You will need a **magnet** in a **plastic baggy** (keep the magnet in the plastic baggy the whole time), a **wire strainer**, a **coffee filter**, a **hotplate**, **water**, **sand**, **sugar**, **marbles**, **iron filings**, **granola**, and two **beakers**. **Looking at the materials and lab we will be using, what are the safety precautions we should take to protect ourselves and materials during the investigation?**

1) Take the mixture of sand, sugar, marbles, iron filings, and granola and use the materials to separate the mixture into its pieces. How will you divide the marbles from the mixture?

2) How will you separate the iron filings from the mixture?

3) How will you separate the granola from the mixture?

4) How will you separate the sand from the mixture?

5) How will you separate the sugar from the mixture?

6) What kind of mixture was this? Explain why.

Simple Middle School Physical Science Investigations   Seven Sides Publishing

Separating Pigments

Directions:

You will need **goggles**, **scissors**, different **pens** and **markers**, an **eyedropper**, **nail polish remover** or **alcohol**, **filter paper** or **chromatography paper**, **test tubes**, a **test tube rack**, **paper clips**, and **rubber stoppers** with holes in them that fit the test tubes. **Looking at the materials and lab we will be using, what are the safety precautions we should take to protect ourselves and materials during the investigation?**

1) You are going to do chromatography today. Take the paper and cut it into strips if you have to, with a point at one end.
2) Make a dot with the pen or marker about an inch from the pointy tip.
3) Take a paper clip and bend it so there is a hook on one end that you poke through the flat end of the strip of paper and put the other end through the hole in the stopper.
4) Take the eyedropper and carefully squirt a little nail polish remover/alcohol into the bottom of a test tube. Then lower the paper with the dot pointy end down into the nail polish remover, but do not let the dot touch it. Fix the stopper at the top of the test tube. Bend the paper clip above the stopper so that the paper does not drop any lower.
5) Repeat the process you did for #s 1-4 for all the different pens and markers you have chosen.
6) Watch the patterns of pigments that separate. The ones that move the fastest go to the top, the ones that move the slowest will be at the bottom.
7) After your pigments have separated, take them out of the test tubes and let them dry out (otherwise, all the stains may go back together again at the top of the paper).
8) Draw pictures of the color bands that came out from your pens and markers on your paper strips below:

Questions:

1) Were these mixtures homogeneous or heterogeneous? Explain why.

2) Did you see the same colors make different patterns?

3) Can the same color of ink be made of different substances?

   a. How did you see this today?

4) How could you use this to find out which pen wrote a note if you needed to?

50 + 50 Does Not Equal 100

Directions and Questions:

You will need **safety goggles**, two **400 mL beakers**, two **100 mL graduated cylinders**, **sand**, **marbles**, 50 mL of **water**, and 50 mL of **rubbing alcohol**. **Looking at the materials and lab we will be using, what are the safety precautions we should take to protect ourselves and materials during the investigation?**

1) Carefully measure 50 mL of water in one graduated cylinder and 50 mL of alcohol in another.

2) Carefully pour the contents of one of the graduated cylinders into the other. What is the total volume now?

3) How do you think this could have happened?

4) Is this a homogeneous mixture or a heterogeneous mixture?

5) Fill a beaker 200 mL full with marbles and another 200 mL full with sand. Gently pour the sand into the marbles, slowly shaking the beaker of the marbles. How many milliliters of marbles and sand is there now in the beaker?

6) Is this a homogeneous mixture or a heterogeneous mixture?

7) Where did the sand go to lose the volume?

8) How can you explain the results for #2 now?

Percent Sugar in Bubble Gum

Directions and Questions:

You will need a **scale** and **bubble gum**. Looking at the materials and lab we will be using, what are the safety precautions we should take to protect ourselves and materials during the investigation?

1) Find the mass of one piece of gum sitting on the wrapper; write this in Data Table 1.
2) Take the gum and chew it. As you are chewing, find the mass of the wrapper and write it in Data Table 1.
3) Subtract the wrapper's mass from the gum and wrapper's mass to get just the unchewed gum's mass; write this in Data Table 1.
4) When you notice no more flavor in the gum, put the chewed gum back on the wrapper and find the chewed gum's mass (do not forget to subtract the wrapper's mass). Write the mass of the chewed gum in Data Table 1.
5) To find the percent of the gum that was not sugar, take the chewed gum's mass and divide it by the unchewed gum's mass times 100.

Data Table 1

Mass of unchewed gum and wrapper	Mass of wrapper	Mass of unchewed gum	Mass of chewed Gum	% of gum, not sugar	% of gum that is sugar
g	g	g	g	g	g

Questions:

1) What was the percentage of gum that is sugar?

2) Where did the sugar and flavor go?

3) So when you are chewing gum, are you technically eating? Explain why.

4) Was this a homogeneous mixture or a heterogeneous mixture? Explain why.

Extension: try this for different brands of gum or even sugarless gum. Show your results below.

Making Ice Cream

Directions:

You will need **gloves**, **milk**, a **spoon**, **crushed ice**, **salt**, **Nesquik** (chocolate or strawberry), a **gallon-size Ziplock bag**, a **quart-size Ziplock bag**, and a **temperature probe** connected to an **interface** that is connected to a **computer** with **Logger Pro**. **Looking at the materials and lab we will be using, what are the safety precautions we should take to protect ourselves and materials during the investigation?**

1) In the quart-sized Ziplock bag, place a cup of milk and two tablespoons of Nesquik. Carefully take as much air out of the bag as possible, seal it shut, and mix the milk and Nesquik.
2) Fill the gallon-size Ziplock bag half full with ice. Place the temperature probe inside the bag of ice to measure the temperature of the ice. Once the temperature stabilizes, write it in Data Table 1 and take the temperature probe out of the bag.
3) Add five tablespoons of salt into the gallon-size Ziplock bag of ice.
4) Place the quart-sized bag into the gallon-sized bag. Remove most of the air in the gallon-sized bag, seal it, put on your gloves, and carefully shake and mix. Be careful not to break the bag on the inside, or you will have salty ice cream, and that will taste nasty. Make sure you keep shaking; otherwise, your ice cream will not be creamy and will freeze into a block of ice (not fun to eat).
5) Once the milk thickens up and turns slushy (it takes 3-5 minutes), open the gallon Ziplock bag and measure the temperature of the ice and salt mixture. Once temperature stabilizes, write this in Data Table 1.
6) Take the quart-sized Ziplock bag out, open your bag, and eat your ice cream with your spoon.

Data Table 1

	Ice by itself	Ice and salt mixture	Temperature Difference
Temperature (C°)	C°	C°	C°

Questions:

1) Why do you think the ice temperature in the gallon-sized bag dropped?

2) Why did the milk mixture freeze to ice cream?

3) From what you observed in this investigation, which water do you think will have a lower freezing point, ocean water, or fresh water in a lake? Explain why.

4) What happened to the temperature of the milk mixture during the investigation? How do you know?

5) How do you think changing the size of the ice cubes or the granules of salt would change the investigation results?

6) Is the ice cream a solid, liquid, or gas? Explain.

The Conductivity of Electrolyte Mixtures

Directions:

You will need a **battery pack**, **batteries**, **solutions** listed in Data Table 1, a **conductivity probe** attached to an **interface** connected to a **computer** with **Logger Pro**, and a **Christmas light** to check the solutions shown in Data Table 1 for conductivity. Having metal in a solution should cause the light bulb to light. **Looking at the materials and lab we will be using, what are the safety precautions we should take to protect ourselves and materials during the investigation?**

1) Connect one wire from the Christmas light to the battery pack with batteries in it. Put the other wires from the battery pack and Christmas light together to see if the light bulb lights. Then dip the free wires into the solutions in the chart below to see if the lightbulb lights. If it lights, the solution conducts electricity; if it does not, the solution does not conduct electricity.
2) Set your conductivity probe to 0-2000 µS/cm). Then measure the conductivity of the substances in the table below, rinsing the probe between each solution.

Data Table 1

Mixture	Light On or Off	Conductivity (µS/cm)
Tap Water (covalent)		
Salt Water (ionic)		
Soap (covalent)		
Orange Juice (both)		
Sugar Water (covalent)		
Soda (both)		
Vinegar (ionic)		
Milk (both)		
Gatorade or Power-aid		

Questions:

1) What pattern do you see about which mixtures conducted electricity or not?

2) Do covalently bonded substances conduct electricity?

3) Do ionically bonded substances conduct electricity?

4) Which bond has metal in it, allowing it to conduct electricity?

5) Gatorade and Power-aid are promoted to replenish electrolytes; does the data show they have many in them?

6) Why do you think we need to replenish electrolytes?

Sugar or Salt

Directions:

You will need a **graduated cylinder**, two **beakers, salt, sugar, water, stirring rods,** and a **scale. Looking at the materials and lab we will be using, what are the safety precautions we should take to protect ourselves and materials during the investigation?**

Part 1:

1) Put 100 mL of water in each beaker. Find the mass of the first beaker with the water. Write that mass in Data Table 1.
2) Add salt to the beaker until no more salt will dissolve. Write the mass of the beaker water and salt in Data Table 1.
3) Subtract the water and beaker from this measurement; this is the mass of the salt. Write this in Data Table 1.
4) Repeat the procedure for # 1-3, doing the same for sugar.

Data Table 1:

Measurements	Salt	Sugar
Mass of beaker and 100mL water (g)		
Mass of beaker, water, and solute (g)		
Mass of solute (g)		

Questions Part 1:

1) Which is more soluble at room temperature, sugar or salt?

2) How much difference is there?

3) How much salt and sugar do you think can dissolve in 10 g of water to make aqueous solutions (aq)?

Directions Part 2:

1) Test it using the same procedures above but with only 10 mL of water in each beaker. Write your data in Data Table 2.

Data Table 2

Measurements	Salt	Sugar
Mass of the beaker and 10mL of water (g)		
Mass of the beaker, water, and solute (g)		
Mass of the solute (g)		

Questions Part 2:

1) How does this compare to your prediction?

2) When we increase a solvent, how does that affect the amount of solute that can be dissolved in an aqueous solution?

The Solubility of Gas in a Liquid

Directions and Questions:

You will need **safety goggles**, two **glass bottles of soda**, two **balloons** (that you fit on the tops of the bottles immediately after they are opened), and a **hotplate. Looking at the materials and lab we will be using, what are the safety precautions we should take to protect ourselves and materials during the investigation?**

1) Open both bottles, fix, and seal a balloon over the opening of each bottle. Place one bottle on the hotplate. Observe both bottles; what do you see happening to the balloons? Why do you think this is happening?

2) Make sure to turn the hotplate off when the experiment is over so it does not boil over. Why do you think we put lids on soda bottles and keep them in the refrigerator?

3) Now take the other bottle, not on the hotplate, and shake it. What do you see happen to the balloon now?

4) Why is this happening?

5) What can you say about how gasses stay dissolved in a liquid solution? (**Hint:** 3 things)

Speed of Dissolving Solutes Lab

Directions and Questions:

You will need **safety goggles**, six **beakers**, cold and warm **water** (cold water could be cooled in the **fridge** overnight or between classes, warm should be heated on a **hotplate**), six **sugar cubes**, a **plastic baggy**, and **beaker tongs. Looking at the materials and lab we will be using, what are the safety precautions we should take to protect ourselves and materials during the investigation?**

1) **Surface Area:** Crush one sugar cube in a plastic baggy. Heat two beakers with equal amounts of water on a hotplate. Just before they boil, drop in the crushed sugar cube in one beaker and a full sugar cube in the other at the same time. Which one dissolves into an aqueous solution (aq) the fastest?

2) **Stirring or not:** Take two beakers with equal amounts of water and place a sugar cube in each at the same time. Stir one beaker and leave the other alone. Which beaker dissolves into an aqueous solution (aq) first?

3) **Hot or Cold:** Take one beaker and heat it on a hotplate. Just before it boils, take a beaker of water out of the fridge. Drop a sugar cube in each at the same time. Which beaker dissolves into an aqueous solution (aq) first?

4) How would you dissolve a substance the fastest way possible?

5) Why were all of these mixtures aqueous solutions (aq)?

Simple Middle School Physical Science Investigations   Seven Sides Publishing

Heat and Saturating Solutions

Directions:

You will need **safety goggles**, a **test tube**, **test tube tongs**, a **test tube rack** (2 or 3 for the whole class), a **hotplate**, a medium-size **beaker of water** that has room for test tubes to sit in safely, **sugar**, and a **solubility curve** for a variety of substances. **Looking at the materials and lab we will be using, what are the safety precautions we should take to protect ourselves and materials during the investigation?**

1) Heat a beaker of water on a hotplate. Heat it but do not let it boil violently.
2) Put some water in a test tube. Stir in sugar until it all dissolves, making an aqueous solution. Then stir in more sugar until no more will dissolve.
3) Place test tube tongs on your test tube and place the test tube in the beaker of hot water. Stir the sugar in the tube; does it dissolve into an aqueous solution (aq)?

4) Put more sugar in the test tube, stirring by holding the tongs and wiggling the test tube back and forth. See that it dissolves. Do this a few times, then place the test tube in a test tube rack and put it into the refrigerator. Check on this at the end of the period or the next day.

Questions:

1) Which water will dissolve more sugar in it, hot or cold?

2) How could you tell?

3) What happened to the sugar solutions you put in the fridge?

4) Was this an aqueous solution (aq)?

5) What is an aqueous solution (aq)?

6) What do you think will happen to the amount of a solute that will dissolve into solvent/water as the temperature of the solvent/water goes up?

7) What do you think will happen to the amount of solute that can dissolve into solvent/water as the temperature of the solvent/water goes down?

8) Find and look at a solubility curve for various substances dissolving in water. Were you correct?

9) What do you think will happen to the amount of solute that can be dissolved in a solvent/water if you double the amount of solvent/water?

Building a Model of a Water Molecule

Directions:

You will need a **balloon**, a **molecular model kit,** and a **Periodic Table. Looking at the materials and lab we will be using, what are the safety precautions we should take to protect ourselves and materials during the investigation?**

1) At the top of your periodic table, label it like this just below:

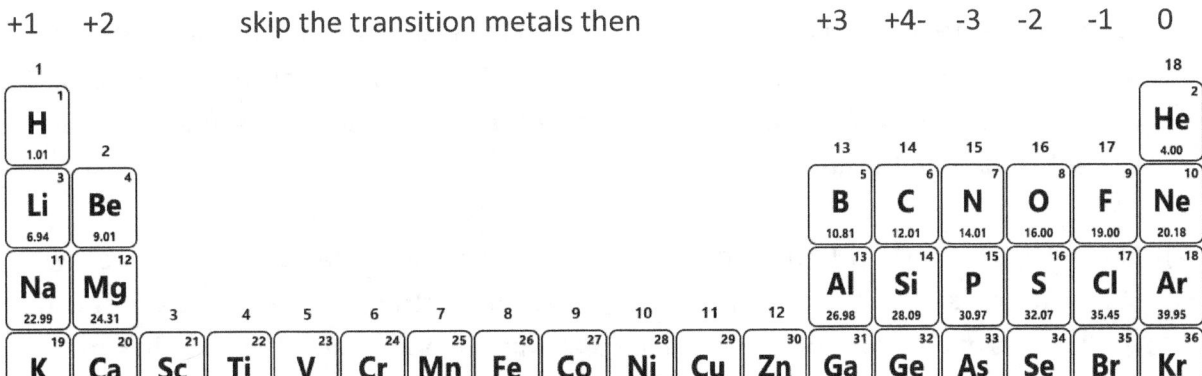

2) Different kits have different colors. In my kit, the:
   a. +1 (one-prong white) represents the Alkali Metals
   b. +2 (two-prong yellow) represents the Alkaline Earth Metals
   c. +3 (three-prong blue) represents the Boron Group
   d. +/- 4 (four-prong black) represents the Carbon Group
   e. -3 (three-prong red) represents the Nitrogen Group
   f. -2 (two-prong blue) represents the Oxygen Group
   g. -1 (one-prong green) represents the Halogens

3) Use the pieces to make two H_2O molecules. The hydrogen side of the molecule is slightly positive, and the oxygen side of the molecule is slightly negative making it polar like a magnet.

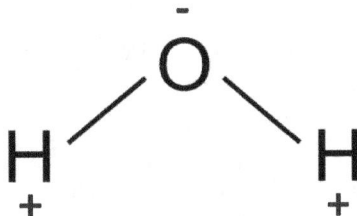

4) Because the water molecule has positive and negative ends, **ions** are attracted to the opposite charges on the water molecule. The positive ions are attracted to the oxygen side, and the negative ions are attracted to the hydrogen side. The same is true for other **polar molecules**; this is why ionic compounds and polar molecules like to dissolve in water. We call water the **universal solvent**.

5) Make a model of liquid water by taking your two water molecules and placing them next to each other where the oxygen of one is sitting between the two hydrogens of the other; this is how water molecules like to stick to each other. The positive ends are attracted to the negative ends; this is why water is **cohesive** (it sticks together).

6) When there are a bunch of them together, they have an equal pull on each other except for the ones on the surface; they are pulled slightly down because they have a slight charge above them. After all, there are no other molecules above them; this is why water has **surface tension**.

7) Since it is charged on both ends, it is also attracted to surfaces like a balloon with a static charge is attracted to a sweater or a wall. This attraction is why see water clinging to the sides of cold cans or glasses of ice tea. This phenomenon is called **adhesion**.
    a. You can model this by taking an inflated balloon, rubbing it on your hair to steal some electrons, and then sticking it to a shirt or wall.

8) You can make a model of ice (solid water) by flipping one of your water molecules and facing the oxygen ends toward each other. When water gets cold, the molecule's charge is not as strong, and the opposite ends are not attracted to each other anymore, so the oxygen atoms come together and share electrons with each other bonding them together. This orientation gives the molecule more space inside it and is why ice floats in liquid water.

Question:

1) Why do you think life depends so much on water?

2) How did these models show the characteristics of water?

3) How were the models inaccurate?

Checking Polarity

Directions:

You will need two **pennies** tails up on a **paper towel**, two **pipettes** or **eyedroppers**, **water**, and **rubbing alcohol. Looking at the materials and lab we will be using, what are the safety precautions we should take to protect ourselves and materials during the investigation?**

1) With a pipet or eyedropper, slowly place drops of water on one penny. Count how many drops you could put on the penny before it spilled off. Write that number in Data Table 1 below.
2) With a different pipet/eyedropper, slowly place drops of rubbing alcohol on the second penny. Count how many drops you could put on the penny before it spilled off. Write that number in Data Table 1.

Data Table 1

Substance Dropped	Number of Drops
Water	
Alcohol	

Questions:

1) Which substance was able to have the most drops put onto the penny?

2) Polarity is in a substance with opposite charges at its ends. Which substance had more polarity?

3) How is a polar molecule like a magnet?

4) **Cohesion** is when water molecules stick to other water molecules. How does the polarity work to allow the water drop to hold together?

   a. Where did we see this in the investigation?

5) **Adhesion** is where water molecules stick to other substances. Where did we see this in the investigation?

6) **Surface tension** is the pulling of the surface water molecules down to the rest of the water because no other force is pulling them up or away. Where did we see surface tension in the investigation?

   a. How do you think this would allow insects to walk on water?

7) How do you think water can defy gravity by moving up the stems of plants?

Celery Transport

Directions and Questions:

You will need fresh **celery** with leaves, a **beaker**, and **red** and **blue food coloring**. This investigation will also work with a **white carnation** and **all colors of food coloring. Looking at the materials and lab we will be using, what are the safety precautions we should take to protect ourselves and materials during the investigation?**

1) In a beaker of water, add a couple of drops of food coloring. Put the celery into the water with the leaves up out of the water. What do you think will happen with the water in the beaker and the celery? Hypothesis:

2) Let the water and celery sit overnight. Come back the next couple of days and observe what you see. What did you see in the celery's leaves?

3) What do you think caused the celery to look like this?

4) How is the xylem in the celery stem like a straw in a drink?

5) Discuss with your teacher and the class how pressure and magnetism were involved with the process you observed in this investigation. Explain what you discussed.

6) Is pressure or magnetism involved with **cohesion** and **adhesion**? Explain.

7) **Transpiration** is the process of water moving up the plant from the roots to the leaves through xylem tubes. How is **cohesion** involved with transpiration?

8) How is **adhesion** involved with transpiration?

Transpiration Pull

Directions and Questions:

You will need a **pressure sensor** and the **tube setup** attached to an **interface** connected to a **computer** with **Logger Pro**. You will then need to find a **plant branch** that will snuggly fit inside the tube, sealing the tube. **Looking at the material and lab we will be using, what are the safety precautions we should take to protect ourselves and materials during the investigation?**

1) Once your pressure sensor is connected to the interface and the computer with the Logger Pro, set the data collection to collect data for 5 minutes. Put the plant branch inside the tube connected to the pressure sensor to seal it.
2) Press "Collect" on the Logger Pro. Watch the data for a few minutes. What do you see happening to the measurement of the pressure sensor?

3) Why do you think this is happening?

4) When the data seems to level out, break the seal, let the pressure equalize with the atmosphere, and stop the data collection if it has not already stopped. Set up the experiment to run again. Did you see the same trend in the results?

5) How does this show evidence of transpiration pull?

Seeing a Stoma

Directions and Questions:

You will need a **textbook, clear scotch tape, lettuce,** a **slide,** a **compound light microscope,** and the **internet** or **textbook. Looking at the materials and lab we will be using, what are the safety precautions we should take to protect ourselves and materials during the investigation?**

1) Take a small piece of scotch tape, put the sticky side on the lettuce, and then peel it off. You should have just removed one layer of cells from the outside of the lettuce. You should see the epidermis, which contains the guard cells and stomata.
2) Place the sticky side of the tape down on the slide. Place the slide on the microscope stage and follow your teacher's instructions on centering and focusing it. Draw a picture of the lettuce epidermis labeling the guard cells and stoma.

3) Research how the guard cells open and close the stoma and describe it here:

4) Why does the rest of the epidermis look like puzzle pieces?

5) What is the function of the epidermis?

6) What is the function of a stoma?

7) How does a stoma allow water to move up the stem?

8) How do cohesion and adhesion help move water up the stem of a plant?

9) What would happen if all the stomata would close in a plant, would transport up the stem be possible? Explain why.

How does Rain Form?

Directions and Questions:

You will need a **glass** or **beaker of ice water. Looking at the materials and lab we will be using,** what are the safety precautions we should take to protect ourselves and materials during the investigation?

1) Why do you see water forming on the outside of the glass?

2) How is this like water forming droplets in the sky, making clouds and rain?

   a. What is the difference between clouds and rain?

3) How is a liquid different from a gas allowing this to happen with water?

4) What allows the water to stick together on the glass?

5) What holds the water drops to the glass?

6) How are **cohesion**, **adhesion**, and **surface tension** involved?

7) If an animal was small enough, explain how it could walk on water.

Virtual Investigations that go with Mixtures and Properties of Water

ExploreLearning.com

    Colligative Properties

    Osmosis

    Diffusion

    Temperature and Particle Motion

    Solubility and Temperature

    Polarity and Intermolecular Forces

    Chemical and Physical Changes STEM Case

    Chemical and Physical Changes Handbook

    Osmosis STEM Case

    Osmosis Handbook

    Properties of Matter STEM Case

    Properties of Matter Handbook

    Water Cycle

    Freezing Point of Salt Water

    Phases of Water

    Phase Changes

    Sticky Molecules

    Solubility and Temperature

    Covalent Bonds

    Polarity and Intermolecular Forces

PhET.colorado.edu

- Concentration
- Gas Properties
- Gas Intro
- Diffusion
- Salts and Solubility
- Sugar and Salt Solutions
- Beer's Law
- Molecule Polarity
- Molecule Shapes
- Molecule Shapes: Basics
- States of Matter
- States of Matter: Basics
- Balloons and Static Electricity

Physicsclassroom.com/Chemistry-Builders/Chemistry

- Classification of Matter
- Molecular Polarity
- Dissociation
- Bond Polarity

Unit 3: Structure of Matter

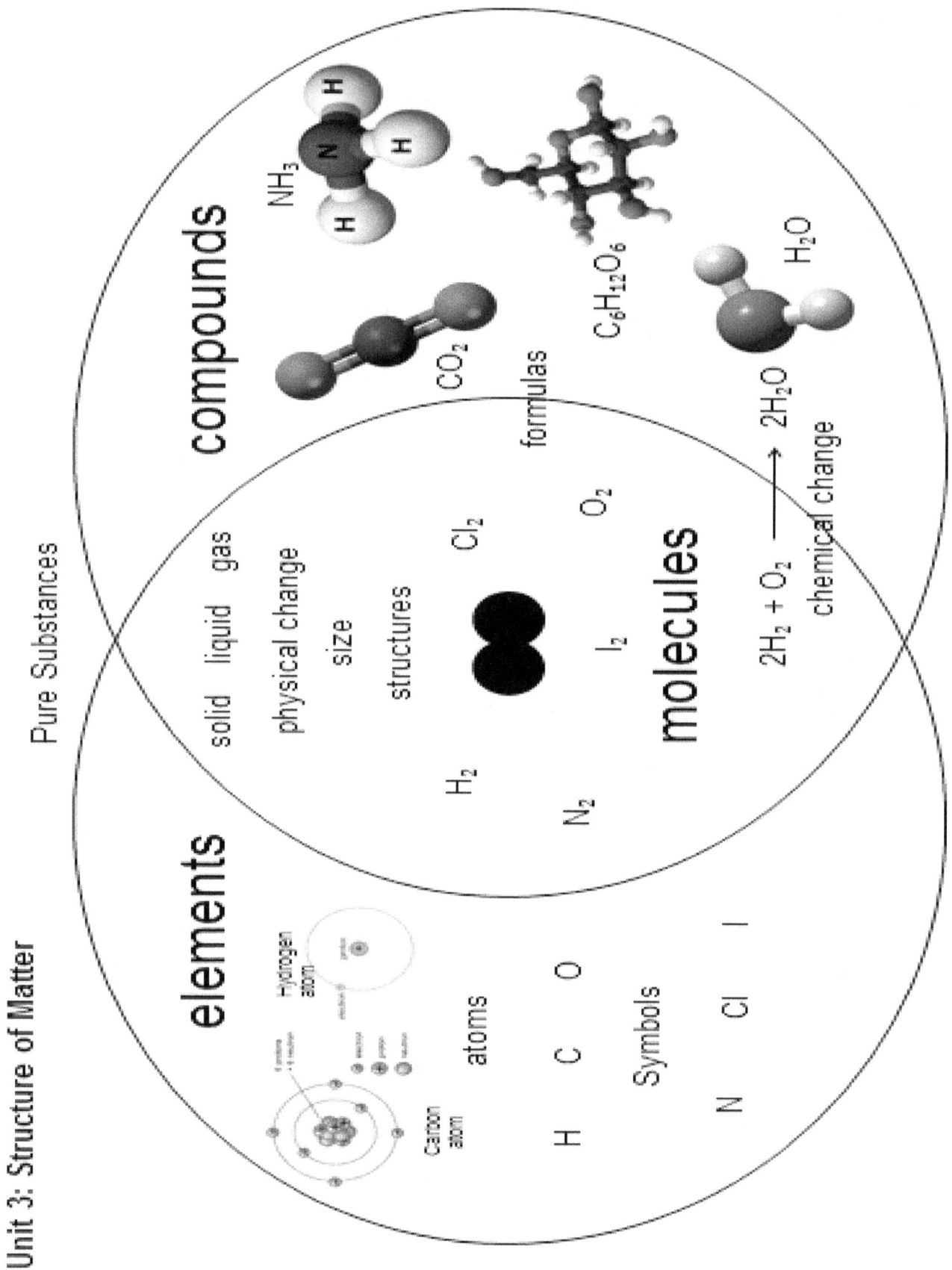

70

Scale Model of a Hydrogen Atom

Directions and Questions:

You will need a **golf ball**, a **bead**, and a **large field** or **parking lot. Looking at the materials and lab we will be using, what are the safety precautions we should take to protect ourselves and materials during this investigation?**

1) Walk out to a large field or parking lot, at least the size of a football field. Keep in mind the space you use still may be too small to be a scale model. You will make a model of a hydrogen atom with 1 proton and 1 electron.
2) On one edge, take a small red bead representing an electron and put it somewhere where you can see it (hang it on a fence or a tiny branch).
3) Walk at least 100 yards away; if you have more room, you can use that. Hold up the golf ball, which is a proton, go through the information and answer the questions that follow. Can you see the bead?

4) This distance is how far away the closest electron speeds around the proton. The speed approaches the speed of light. It moves so fast that it makes a ball the size of a football stadium. If you have ever seen a fan moving fast, does it look like a disk? But is it a disk?

    a. So we have to ask ourselves, the atoms, the pixels of our universe, look like solid balls (see the picture below), but are they balls? Explain why.

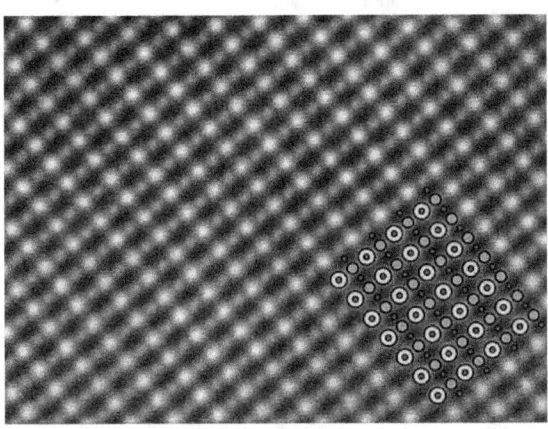

Atom resolution STEM image by Magnunor (Own work) CC BY-SA 4.0.

The illusion of the electron is a similar type of illusion. When an object moves close to the speed of light, time stops for that object. Quantum physics allows things without time to be everywhere they could typically be at the same time. So, why is the electron said to be everywhere in the electron cloud simultaneously?

5) What do you see between the proton and electron?

6) Quarks orbit inside protons, making them appear solid. Is there anything solid inside the atom?

7) If atoms are mostly empty space, why do they look solid?

8) What would happen to the electron if time were to stop?

9) If that would happen to the electron, what would happen to the atom?

10) What would happen to all atoms in the area where time stops?

11) This answer is why we call this the space-time continuum. You cannot have space without time; this is how black holes form. When time disappears, so does space. What color is a black hole? Explain why.

12) This model also shows that the pixel of our universe is an illusion. If the pixel of our universe is an illusion, what does that say about our universe?

13) Why do we consider the atomic theory a theory? Why is it not a law or hypothesis?

Simple Middle School Physical Science Investigations   Seven Sides Publishing

Model of an Atom Showing the Illusion

Take a **golf ball** and **small bead** out to a **large parking lot** or **field** larger than a football/soccer field. Place the small bead (electron) on the edge of the field/parking lot. Walk to the center of the field/lot with the golf ball (proton). The electron spins around the proton this far away, so fast that it makes it look like a ball about the size of a football stadium.

1) What is between the electron and proton?

2) What happens to the atom if the electron stops moving?

3) What happens to the electron if time stops?

4) What happens to the atom if time stops?

5) What happens if you stop the electron motion?

6) What happens to the part of the universe where you stop time?

This answer is why time and space are connected in space-time. You cannot have one without the other.

7) The whole universe has and can fit inside of what structure?

   a. We know all this is possible because the atom is mostly made up of what?

$E=mc^2$ says matter can be converted into energy, and energy can be converted into matter. The way the atom is structured shows us this is not only possible, but it is real.

Simple Middle School Physical Science Investigations   Seven Sides Publishing

Building Bohr Models

Directions:

You will need a **film case** filled with **three colors of beads** (I use red, white, and blue) and a **periodic table. Looking at the materials and lab we will be using, what are the safety precautions we should take to protect ourselves and materials during the investigation?**

1) Designate which color bead represents which part of the atom. Here is an example: Red beads are electrons, white beads are neutrons, and blue beads are protons. I use these colors because this is how my periodic table was colored. These colors help my students learn how and why the periodic table works.

2) Carefully empty the film case. As the students empty those on the table, some beads will bounce out. I call this radiation because radiation is particles and energy coming out of an atom.

3) Also know that a neutron is the combination of a proton and an electron. This is why it is neutral. Proton (p+) + Electron (e-) = Neutron (n°) or 1+ + 1- = 0.

4) Place the empty film canister in the center circle labeled the nucleus on page 76.

5) Now make a **Hydrogen atom** by looking at the periodic table and having the student look at the atomic number; this tells us how many protons are in that atom. What defines the element is the number of protons. The atomic number for Hydrogen is 1. So place one blue proton in the nucleus (film case). The number of protons also tells you how many electrons there will be in a neutral atom. So place one red bead in the first orbit closest to the nucleus. Protons and neutrons have mass, and electrons do not. Look at the average atomic mass at the bottom of the element's box. Round it to the closest whole number; this tells you the mass of most of the atoms of that element. Since the mass is one and we have one bead in the nucleus, we have completed the hydrogen 1 (H-1) isotope.

6) Add a neutron (white bead) to the nucleus; this makes a Hydrogen 2 (H-2) isotope. One proton and one neutron in the nucleus (2 beads) give us a mass of 2. Hydrogen 2 isotope is another version of the same element.

7) Next, make a **Helium atom**. Look at the periodic table and find Helium. How many protons does it have (look at the atomic number)? Add a blue bead to the nucleus to give it 2 protons.

8) How many electrons does Helium need to have? Add this to the first circle but on the opposite side from the other electron.

9) Why would the electron be there?

10) Now, look at the mass of the Helium. How many does it have (remember to round to the whole number)?

11) How many beads are in your nucleus?

12) Add two white beads in the nucleus (film canister) because the average mass is 4; this makes a Helium 4 (He-4) isotope.

13) Now take one neutron (white bead) out to make Helium 3 (He-3) isotope; this is what is on the moon. Scientists are going to want this in the future to cause fusion reactions to produce electricity.

14) We have now filled the first energy level of electrons. Notice we are on the right end of the periodic table. No more electrons can go into this circle.

15) We will now make a **Lithium atom**. How many protons does it have? Place a blue bead in the nucleus (film canister) until you match that number of protons.

16) How many electrons does Lithium have? Place an electron (red bead) on the second circle. We are now at the second energy level.

17) Look at the average mass of Lithium. Add neutrons (white beads) to the nucleus until you have matched the mass of Lithium.

18) Now go and make the other isotopes, **Carbon 12** (which is in all life) and **Carbon 14** (which is radioactive and starts to decay when an organism dies), which are essential to life. **Carbon 14** turns into **Nitrogen** by taking an electron out of a neutron (a proton and an electron); this leaves a proton in the nucleus, making it a Nitrogen atom.

19) Then build **Fluorine** and **Neon**. Neon fills the second energy level, so no more electrons can fit here to make the next elements. Where do you think they will go?

20) Now build **Sodium**, **Magnesium**, and **Potassium**.

Bohr Model

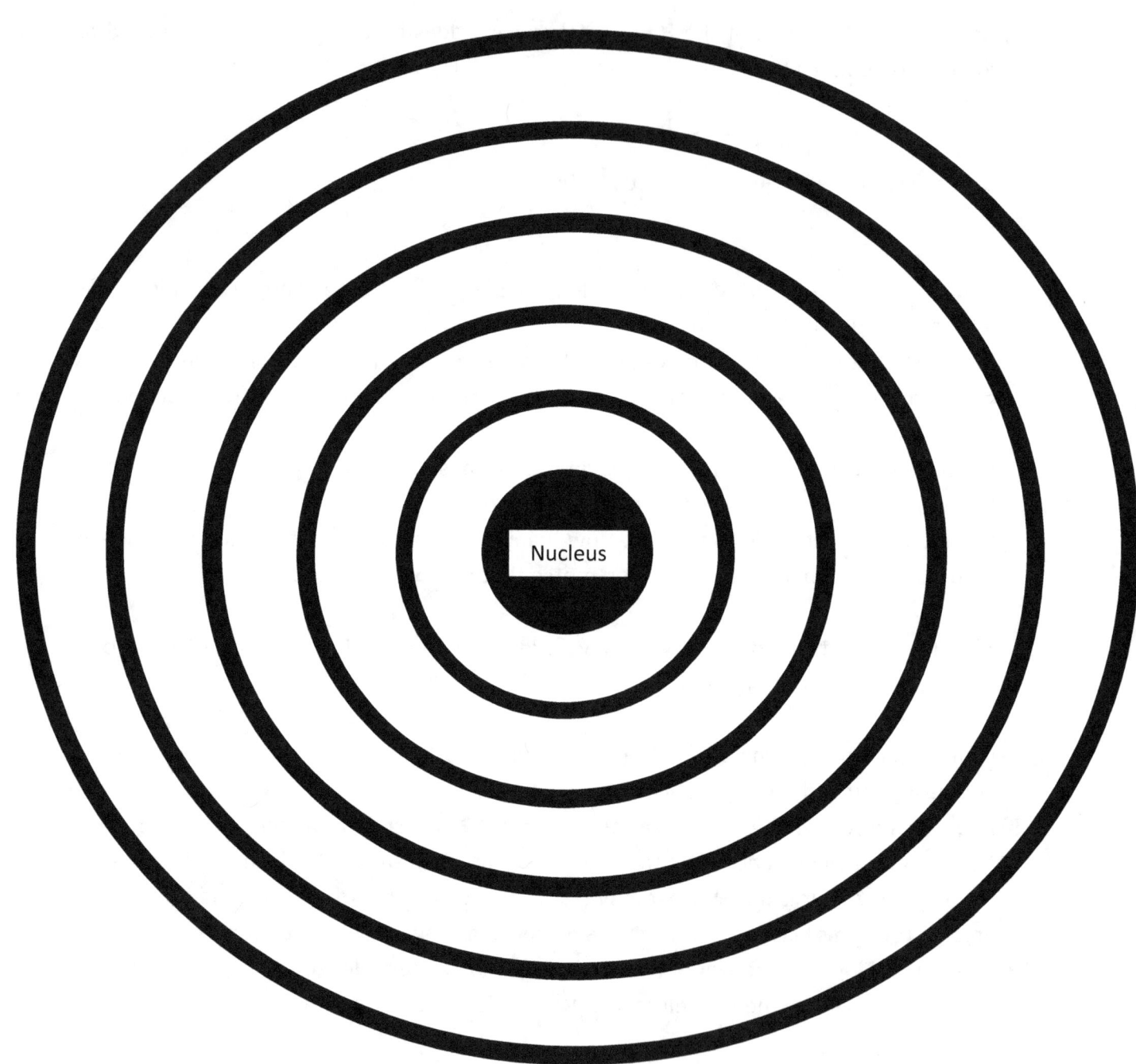

Questions:

1) How is the atomic mass determined?

2) How do you find out the number of protons for an element?

3) How do you find out the number of neutrons for an element?

4) How do you find out the number of electrons for each element?

5) How many electrons can go into the first energy level?

6) How many electrons can go into the second energy level?

7) How is the periodic table structured to tell us about the atoms of each element?

8) How does this model not accurately show where electrons orbit in atoms?

Simple Middle School Physical Science Investigations   Seven Sides Publishing

Finding the Period in the Periodic Table

Directions:

You will need a **periodic table**. Use the periodic table to fill in the Data Tables and answer the questions below.

Data Table 1

Element	Total # of Electrons	# of Level 1 Electrons	# of Level 2 Electrons	# of Level 3 Electrons
Magnesium				
Carbon				
Aluminum				
Sodium				
Florine				
Helium				
Oxygen				
Argon				
Silicon				
Nitrogen				

Questions for Data Table 1:

1) How many electrons can fill the first energy level?

2) How many elements do not fill that first energy level?

3) How many electrons fill the second energy level?

Data Table 2

Element	Energy Level of Outer Electrons	Located in period	Number of Outer Electrons	Located in Group
Magnesium				
Carbon				
Aluminum				
Sodium				
Fluorine				
Helium				
Oxygen				
Argon				
Silicon				
Nitrogen				

Questions Data Table 2:

1) How do you know an element's group?

2) Which elements have full outer shells?

3) What does the element's period tell us about the energy levels they have?

4) How does the periodic table seem to be organized?

5) How are elements different from compounds?

Metal or Nonmetal

Directions:

You will need samples of **sulfur, charcoal, copper, aluminum foil, mechanical pencil lead** (graphite), a **battery pack, batteries,** and a **Christmas light** with the ends of the wires stripped of insulation. **Looking at the materials and lab we will be using, what are the safety precautions we should take to protect ourselves and materials during the investigation?**

1) **Metals** are shiny, malleable, ductile, and conduct heat and electricity well. **Nonmetals** are brittle, dull, and do not conduct heat and electricity well.
2) Observe the five samples and fill in Data Table 1 below on those materials. Use the battery pack with batteries and the Christmas light to see if the materials conduct enough electricity to light the Christmas light when put into a circuit.

Data Table 1

Sample	Shiny or Dull	Brittle or Malleable	Conduct Electricity?	Metal or Nonmetal?
Sulfur				
Charcoal				
Copper				
Aluminum Foil				
Mechanical Pencil Lead				

Questions:

1) What are the characteristics of metals?

2) What are the characteristics of nonmetals?

3) Look at the periodic table and find carbon. Why do you think graphite conducted electricity?

Periodic Table Activity

Using your **Periodic Table**, color the following with **colored pencils**:

1) Nonmetals are (yellow)
2) Metalloids are (green)
3) Metals are (blue)
4) Outline the alkali metals in (red)
5) Outline the alkaline earth metals in (black)
6) Outline the transition metals in (brown)
7) Outline the halogens in (blue)
8) Outline the noble gasses in (purple)

Add the following to the Periodic Table:

1) Atomic numbers
2) Group/family numbers
3) Periods/energy levels
4) Oxidation numbers
5) Label the Lanthanide series and Actinide series

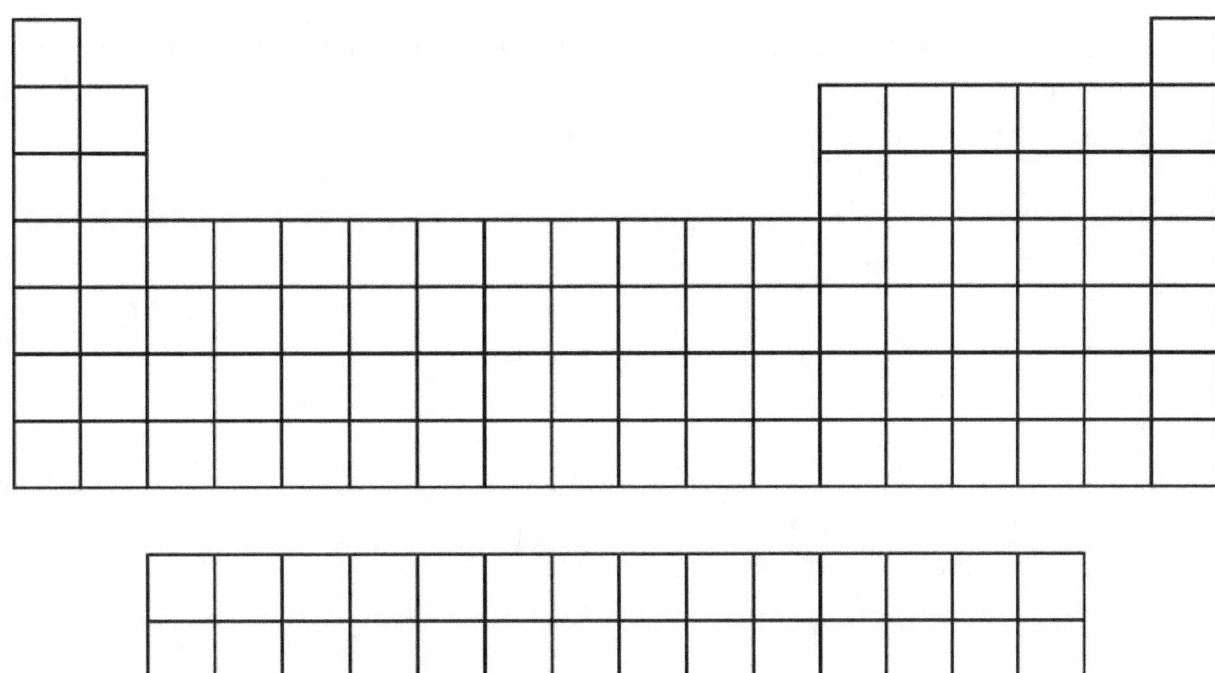

Questions:

1) What do groups/families have in common?

2) What do periods have in common?

3) Which group/family has full outer shells?

4) Where are the metals on the periodic table?

5) Where are the nonmetals on the periodic table?

6) Where are the metalloids on the periodic table?

7) If you can't remember from previous investigations, research the internet to find these three things:
    a. What are the characteristics of metals?

    b. What are the characteristics of nonmetals?

    c. What are the characteristics of metalloids?

Making a Graphite Light Bulb (A)

Directions:

You will need **.2 to .5 mm graphite mechanical pencil lead** (the thinner, the better because it causes more resistance), a **glass jar** with a **lid**, three **wires with alligator clips**, two **6 volt lantern batteries**, and **blue tac** (used to fix papers and posters to walls). **Looking at the materials and lab we will be using, what are the safety precautions we should take to protect ourselves and materials during the investigation?**

1) Take two pieces of the blue tac and fix them to the inside lid of the jar. Place an alligator clip from both wires on the blue tac to hold the mouths up.
2) Take the mechanical pencil lead (graphite) and break a piece off big enough to fit across the two alligator clips and inside the jar's lid. Place the glass jar over the top of the lid covering the graphite and alligator clips.
3) Turn off the lights in the room. Take the other ends of the wires and place one on the "+" end of the first battery and the other wire on the "−" end of the second battery. To complete the circuit, take the third wire with alligator clips and place one clip on the "−" end of the first battery and the other clip on the "+" end of the second battery.
4) Make sure to disconnect the clips from the battery setup when you have finished.

Questions:

1) Why do you think the graphite lit up?

2) Look at the periodic table and find Carbon; this makes graphite. Why do you think this nonmetal was able to be used here?

3) Is Carbon a conductor or insulator? How do you know?

4) When you slow the flow of electrons in graphite, what do you see given off?

5) How do you think an incandescent light bulb lights up when a current flows through it?

6) Explain how electrons and photons are involved with what is happening inside this light bulb.

7) Why did we cover the graphite with the glass jar?

Simple Middle School Physical Science Investigations   Seven Sides Publishing

Making Models of Compounds

Directions and Question:

You will need a **molecular model kit** and a **Periodic Table**. Looking at the materials and lab we will be using, what are the safety precautions we should take to protect ourselves and materials during the investigation?

1) At the top of your periodic table, label it like this just below:

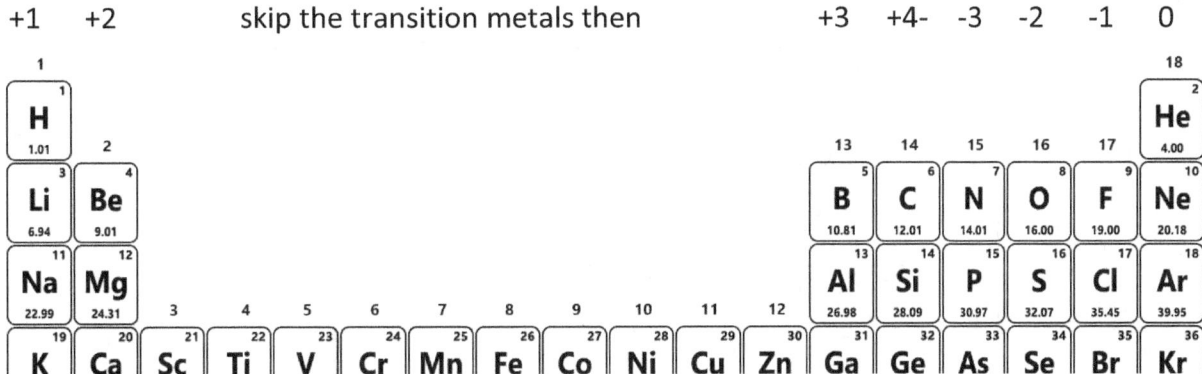

2) Different kits have different colors. In my kit, the:
   a. +1 (one-prong white) represents the Alkali Metals
   b. +2 (two-prong yellow) represents the Alkaline Earth Metals
   c. +3 (three-prong blue) represents the Boron Group
   d. +/- 4 (four-prong black) represents the Carbon Group
   e. -3 (three-prong red) represents the Nitrogen Group
   f. -2 (two-prong blue) represents the Oxygen Group
   g. -1 (one-prong green) represents the Halogens

3) **Compounds** are two or more elements chemically combined together. **Molecules** are two or more atoms combined together to form a substance.

4) The different pieces in #2 represent the elements in those groups. Put the following ionic compounds together:

   KF MgI_2 BeS Na_2O $AlBr_3$ CH_3Cl NHF_2

5) How many elements and atoms are in each formula?

Formula	Number of Elements	Number of Atoms
KF		
MgI_2		
BeS		
Na_2O		
$AlBr_3$		
CH_3Cl		
NHF_2		
CrF_2		
Li_3PO_4		
NaOH		
Co_3N_2		
$MgBr_2$		
CaO		
$(NH_4)_2SO_3$		
$Al_2(SO_4)_3$		
Na_3P		

6) Which of these compounds are not molecules?

Making Molecular Models

Directions and Questions:

You will need a **molecular model kit** and a **Periodic Table**. Looking at the materials and lab we will be using, what are the safety precautions we should take to protect ourselves and materials during the investigation?

1) At the top of your periodic table, label it like this just below:

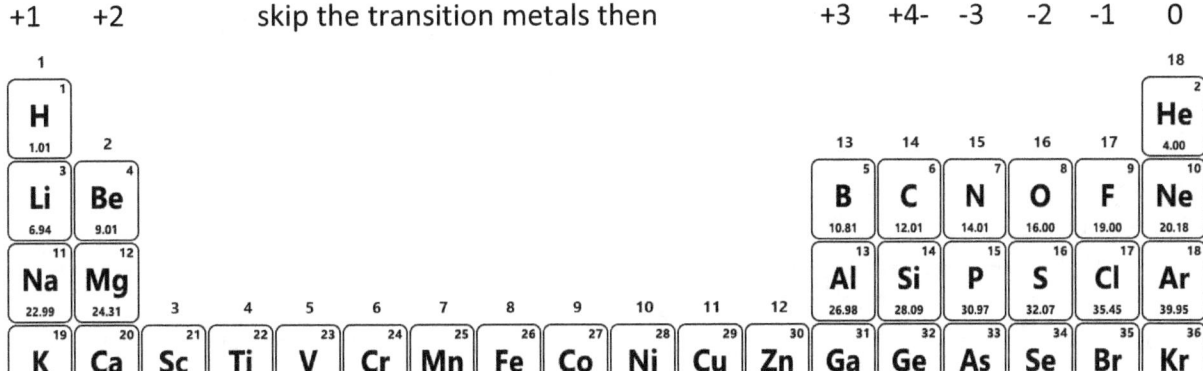

2) Different kits have different colors. In my kit, the:

   a. +1 (one-prong white) represents the Alkali Metals
   b. +2 (two-prong yellow) represents the Alkaline Earth Metals
   c. +3 (three-prong blue) represents the Boron Group
   d. +/- 4 (four-prong black) represents the Carbon Group
   e. -3 (three-prong red) represents the Nitrogen Group
   f. -2 (two-prong blue) represents the Oxygen Group
   g. -1 (one-prong green) represents the Halogens

3) Molecules are two or more atoms combined together to form a substance.
4) The different pieces in #2 represent the elements in those groups. Put the following molecules together:

H_2O NH_3 $SiCl_4$ I_2 SCl_2 O_2 AsH_3

5) Count how many elements and atoms there are in each molecule.

Formula	Number of Elements	Number of Atoms
H_2O		
NH_3		
$SiCl_4$		
I_2		
SCl_2		
O_2		
AsH_3		
P_4S_5		
H_2		
SF_6		
Si_2Br_6		
CH_4		
B_2Si		
N_2		
Cl_2		
N_2O_5		

6) Which of these molecules are not compounds?

Virtual Investigations that go with Structure of Matter

ExploreLearning.com

    Chemical Changes

    Ionic Bonds

    Covalent Bonds

    Element Builder

    Electron Configuration

    Polarity and Intermolecular Forces

    Collision Theory

    Properties of Matter STEM Case

    Properties of Matter Handbook

    Physical and Chemical Changes STEM Case

    Physical and Chemical Changes Handbook

    Electrons and Chemical Reactions STEM Case

    Electrons and Chemical Reactions Handbook

PhET.colorado.edu:

    Build an Atom

    Build a Molecule

    Molecule Polarity

    Molecule Shapes

    Molecule Shapes: Basics

Physicsclassroom.com/Concept-Builders/Chemistry

    Classification of Matter

    Name that Element

- Particles... Words... Formulas
- Formulas and Atom Counting
- Chemical Equations
- Atomic Models
- Subatomic Particles
- Isotopes
- Periodic Table Battleship
- Periodic Trends
- Ionic Bonding
- Bond Polarity
- Molecular Polarity
- Collision Model of Reaction Rates

Unit 4: Chemical Reactions

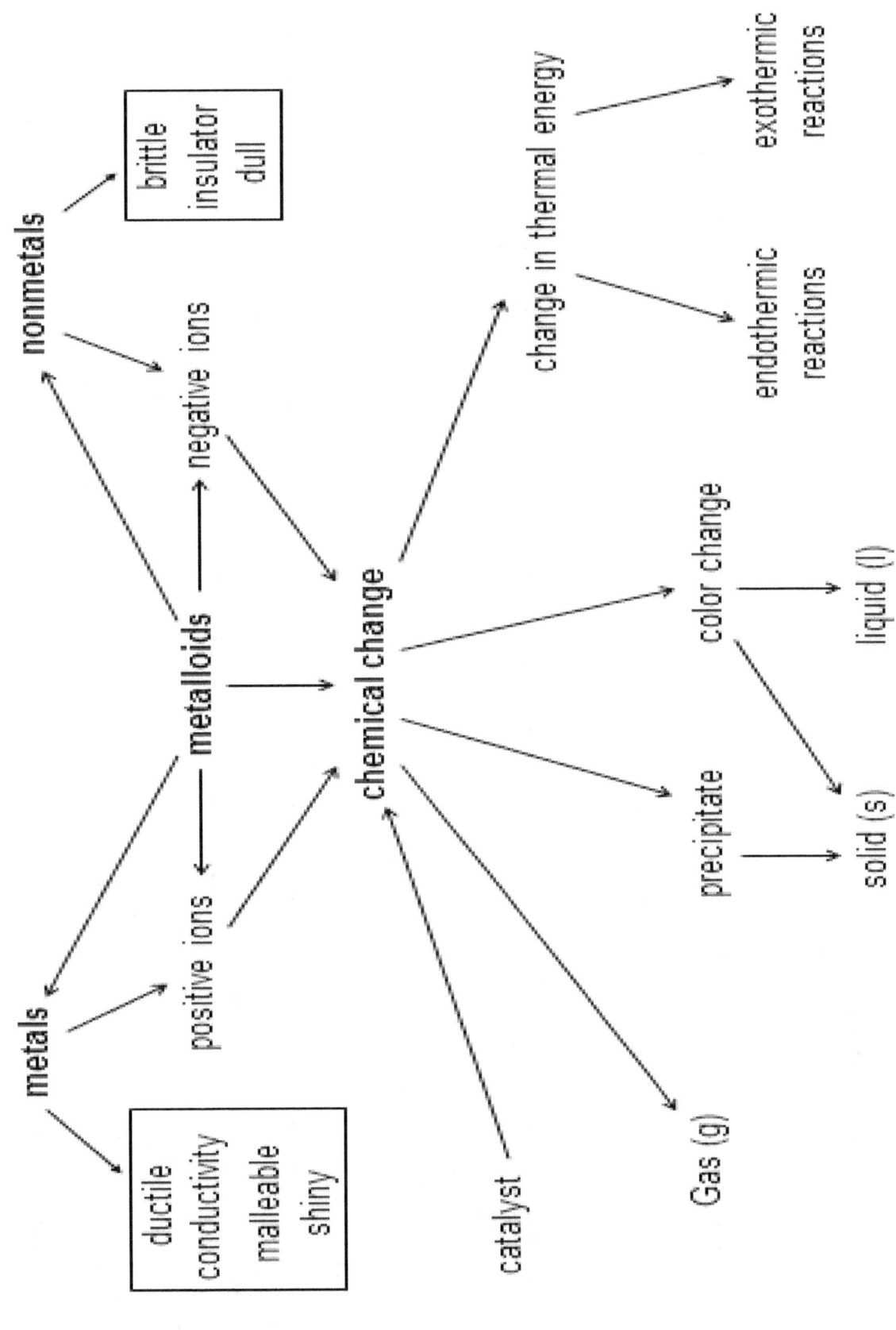

Unit 4: Chemical Reactions

synthesis reactions

single-replacement reactions

endothermic reactions

decomposition reactions

double-replacement reactions

exothermic reactions

catalyst:

inhibitor:

```
oxidation
number  ──→  balanced chemical equation
             ↑
          coefficients
             
reactants  ──▶  products

solid (s)
liquid (l)
aqueous solution (aq)
gas (g)
```

Nuclear Fission:

Nuclear Fusion:

Temperature and Reaction Rates

Directions:

You will need **safety goggles**, a **beaker** of **cold water** cooled by a **refrigerator**, a **beaker** of **warm water** heated by a **hotplate**, and two **Alka-Seltzer tablets**. **Looking at the materials and lab we will be using, what are the safety precautions we should take to protect ourselves and materials during the investigation?**

1) Heat one beaker with a hotplate to the point just before it boils.

2) Then take the other beaker out of the refrigerator.

3) At the same time, add an Alka-Seltzer tablet to each of the beakers and see which tablet finishes reacting first.

Questions:

1) What showed us a chemical change took place?

2) Which water made the tablet react the fastest?

3) Why do you think this happened?

Observing a Catalyst

Directions and Questions:

You will need **safety goggles**, an **apron**, a small bottle of **soda** (diet has more fizz), and a **Mentos tablet**. Looking at the materials and lab we will be using, what are the safety precautions we should take to protect ourselves and materials during the investigation?

1) Open the bottle of soda. What do you observe happening?

2) Now drop the Mentos (the catalyst) into the bottle of soda. What do you observe now?

3) What indicated a chemical reaction took place?

4) This reaction quickly converted all the carbonic acid in the soda to carbon dioxide and water. The Mentos catalyst will keep working until all the carbonic acid is gone without using up the catalyst. Try to devise an experiment showing the catalyst does not go away.

Change in Temperature in Chemical Reactions

Directions and Questions:

You will need **safety goggles**, a **beaker**, **water**, **Epsom salt**, **Borax**, **laundry detergent**, **baking soda**, **vinegar**, **matches**, and a **temperature probe** attached to an **interface** connected to a **computer** with **Logger Pro**. **Looking at the materials and lab we will be using, what are the safety precautions we should take to protect ourselves and materials during the investigation?**

1) **Endothermic Reactions** absorb energy into the reactions making the measured temperature drop. **Exothermic reactions** release energy into the environment making the measured temperature go up. We will conduct a few simple reactions and measure with a temperature probe whether the temperature goes up or down and thus classify those reactions as endothermic or exothermic.

2) Put a temperature probe into a beaker of water. Once the temperature stabilizes, add a small spoon full of Epsom salt to it. Gently stir the mixture slowly, watching the temperature readings on Logger Pro. What happened to the temperature during this reaction?

    a. Was this an endothermic or exothermic reaction? How do you know?

3) Rinse everything off and put water into your beaker again. Put a temperature probe into the beaker of water. Once the temperature stabilizes, add a small spoonful of borax to it. Gently stir the mixture with the temperature probe as it reacts, watching the Logger Pro temperature reading. What happened to the temperature during this reaction?

    a. Was this an endothermic or exothermic reaction? How do you know?

4) Rinse everything off and put water into your beaker again. Put a temperature probe into the water. Once the temperature stabilizes, add laundry detergent. Gently stir the mixture as it reacts, watching the temperature reading on the Logger Pro. What happened to the temperature during the reaction?

a. Was this an endothermic or exothermic reaction? How do you know?

5) Rinse everything off and put water into your beaker again. Put a temperature probe into the water. Once the temperature stabilizes, add baking soda to it. Gently stir the mixture as it reacts, watching the temperature reading on the Logger Pro. What happened to the temperature during the reaction?

a. Was this an endothermic or exothermic reaction? How do you know?

6) Rinse everything off and put vinegar into your beaker this time. Put the temperature probe into the vinegar. Once the temperature stabilizes, add baking soda to it. Gently stir the mixture as it reacts, watching the temperature readings on the Logger Pro. What happened to the temperature during the reaction?

a. Was this an endothermic or exothermic reaction? How do you know?

7) Light a match and watch the reaction that takes place with the resulting fire. Is this an endothermic or exothermic reaction? How do you know?

8) When fireworks go off, is this an endothermic or exothermic reaction? How do you know?

9) Instant cold packs are made up of two bags, one inside the other. One bag contains water, while the other holds a chemical like calcium ammonium nitrate. When you hit it and shake it, the bag on the inside breaks mixing the two contents causing the temperature to drop; is this an endothermic or exothermic reaction? How do you know?

10) What were we measuring each time that told us a chemical reaction took place?

Removing Carbon from Sugar

Directions and Questions:

Perform this reaction under a **fume hood** or **go outside**. Some unhealthy gasses and heat are produced from this reaction, so once it has started, stand back and watch but do not touch. You will need **safety goggles**, an **apron**, **sugar**, **baking soda**, **lighter fluid**, a **ceramic bowl with sand**, and a **long-necked lighter**. **Looking at the materials and lab we will be using, what are the safety precautions we should take to protect ourselves and materials during the investigation?**

1) Now take the ceramic bowl of sand and have your teacher spray some lighter fluid on it. Place a mixture of 40 g sugar and 10 g baking soda on it. Make sure there is nothing near the bowl. Have a fire extinguisher ready just in case the fire gets out of hand. Then with a long-neck lighter, light it on fire. What do you see happening?

2) The sugar reacts with the oxygen producing carbon dioxide and water. When the oxygen runs out, the sugar breaks down to solid carbon (what looks like a black snake) and water vapor. The baking soda changed to carbon dioxide, water, and sodium carbonate (this is what is stealing the oxygen to help the sugar create a black snake). Write the word equation for the combined reactions taking place.

3) Balance the reaction taking place.

   $NaHCO_3$ + $C_{12}H_{22}O_{11}$ + O_2 → CO_2 + H_2O + Na_2CO_3

4) Since this is burning sugar like life does to live, is respiration a physical or chemical change?

Wait until it cools before touching it. The black snake will stain whatever touches it, so be careful as you dispose of it using your teacher's instructions.

Conservation of Mass in Equations

Directions:

Draw lines from the elements in the reactants (the left side of the arrow) to the elements in the products (the right side of the arrow) for each balanced chemical equation to show how elements do not go away in a chemical reaction; they just get rearranged. An example is done below.

Example: $2H_2 + CO \rightarrow CH_3OH$

1) $2H_2O \rightarrow 2H_2 + O_2$

2) $HCl + NaHCO_3 \rightarrow CO_2 + H_2O + NaCl$

3) $C_6H_{12}O_6 + 6O_2 \rightarrow 6CO_2 + 6H_2O$

4) $CH_4 + 2O_2 \rightarrow CO_2 + 2H_2O$

5) $2H_2 + 2O_2 \rightarrow 2H_2O$

6) $C_2H_5OH + 2O_2 \rightarrow 2CO_2 + 3H_2O$

7) $6CO_2 + 6H_2O \rightarrow C_6H_{12}O_6 + 6O_2$

8) $2LiOH + CO_2 \rightarrow Li_2CO_3 + H_2O$

9) $(NH_4)_2Cr_2O_7 \rightarrow Cr_2O_3 + N_2 + 4H_2O$

10) $4NH_3 + 5O_2 \rightarrow 4NO + 6H_2O$

Questions:

1) How do these equations show mass is conserved?

2) How can a reaction have different reactants from its products but still conserve mass?

3) What do the numbers in front of the formulas represent?

4) How can you tell how many elements there are in an equation?

5) Which careers use the information in the conservation of mass?

Simple Middle School Physical Science InvestigationsSeven Sides Publishing

Home Chemistry

Directions and Questions:

You will need **safety goggles, liver, hydrogen peroxide**, three small **cups, water, vinegar**, and **baking soda**. Looking at the materials and lab we will be using, what are the safety precautions we should take to protect ourselves and materials during the investigation?

1) All living things produce an enzyme called catalase. Catalase helps break down hydrogen peroxide into water and oxygen. Place a piece of liver in a cup and pour some hydrogen peroxide we use as an antiseptic on it. What do you observe?

   a. Write a balanced equation for the reaction.

2) One of the most useful natural gasses we use as fuel is methane (CH_4). It combines with oxygen in a combustion reaction to create heat, carbon dioxide, and water. Write a balanced equation for the reaction.

3) Place baking soda ($NaHCO_3$) and vinegar (CH_3COOH) together in a cup and watch the reaction. The products are carbon dioxide (CO_2), water (H_2O), and sodium acetate (CH_3COONa). What do you observe?

   a. Write a balanced equation for this reaction.

4) What do you think we use baking soda for when we cook?

Types of Chemical Reactions

Equipment and Safety:

You will need **safety goggles**, an **apron**, a **small aluminum pan**, a **digital scale**, **steel wool**, **baking soda**, **matches**, a **copper sulfate solution** in a **beaker**, a **nail** or **screw**, a **lighter**, a **small beaker**, **beaker tongs**, a **test tube**, and a **test tube holder**. Looking at the materials and lab we will be using, what are the safety precautions we should take to protect ourselves and materials during the investigation?

Prep for Reaction 3

1) Take the **nail or screw** and write down how it appears now.

2) Stand it up in the small amount of **copper sulfate** solution and come back and look at it after the other reactions are done.

Reaction 1

3) Place the aluminum pan on the digital scale and zero it out.
4) Take some **steel wool** and place it in a pan on a digital scale. What is the mass of the steel wool?

5) What is the color of the steel wool before burning?

6) Take the lighter and light the steel wool. Place the lighter's flame into the steel wool and observe the reaction. What do you see happening during the reaction?

7) Notice the color change. What is the color of the burned steel wool?

8) What is the mass of the burnt steel wool?

9) Why do you think the mass changed?

10) This reaction was a combustion reaction where the iron in the steel wool combined with oxygen in the air, using fire, forming the iron oxide. Write a word equation of the reaction.

11) Balance the equation below for this chemical reaction.

$$Fe + O_2 \rightarrow Fe_2O_3$$

Reaction 2

12) Take some **baking soda** ($NaHCO_3$) and place it in a test tube; make sure you hold the test tube with the test tube holder. Take the lighter and heat the test tube with the baking soda in it. What do you see forming on the inside of the glass of the test tube?

13) Now light a match and place it inside the mouth of the test tube. What happens to the flame?

14) What color is the solid at the bottom of the test tube?

15) What you saw was water form on the inside of the test tube; and carbon dioxide gas form that snuffed the flame of the match out. Sodium carbonate is what is left. Write a word equation for the reaction.

16) Balance the chemical equation below for this reaction.

$$2 NaHCO_3 \rightarrow CO_2 + H_2O + Na_2CO_3$$

Reaction 3

17) Go back to your **nail or screw** and carefully pull it out of the liquid. How does it look now?

18) The iron and the copper switched places. The iron is now combined with the sulfate, and the copper is by itself. Write a word equation for the reaction.

19) Balance the equation for this reaction.

$$2 Fe + 3 CuSO_4 \rightarrow Fe_2(SO_4)_3 + 3 Cu$$

Acid-Base Reaction

20) Write a word equation for hydrochloric acid mixing with sodium hydroxide that makes water and table salt.

21) Balance the equation for this reaction.

$$HCl + NaOH \rightarrow H_2O + NaCl$$

Conservation of Mass

Directions and Questions:

You will need a **scale**, a **Ziploc bag**, **baking soda**, **vinegar**, and a **disposable pipette**. **Looking at the materials and lab we will be using, what are the safety precautions we should take to protect ourselves and materials during the investigation?**

1) Take a small Ziploc bag and put a spoonful of baking powder in it. Suck up some vinegar into the pipette and place the pipette in the bag. Seal the bag shut.
2) Find the mass of the bag and its contents before the reaction. What is the mass?

3) Squeeze the vinegar from the pipette into the baking soda. What do you observe?

4) Find the mass of the bag and its contents after the reaction. What is the mass?

5) Why do you think we made the reaction take place inside a sealed bag?

6) What did you see and feel that let you know a chemical reaction took place?

7) Was it an endothermic or exothermic reaction? How did you know?

8) Compare the mass of the bag and its contents before the reaction and after.

9) Why do you think the results come out as they did?

10) Use your textbook or the internet to find how the conservation of mass is stated, and write it here.

11) How did this reaction show the conservation of mass?

12) This reaction also showed the conservation of energy. How did this show energy change through the reaction?

The Law of Conservation of Mass

Directions and Questions:

You will need **safety goggles**, an **apron**, a **graduated cylinder**, 10 mL of **potassium iodide solution** in a **250 mL Erlenmeyer flask**, a **test tube** half full with a **lead nitrate solution** placed in the Erlenmeyer flask so that it does not mix with the potassium iodide, and a **rubber stopper** sealing the flask. **Looking at the materials and lab we will be using, what are the safety precautions we should take to protect ourselves and materials during the investigation?**

1) Measure the mass of the Erlenmeyer flask and its contents. What is it?

2) Carefully and slowly invert the flask, letting the test tube's contents mix with the flask's contents while keeping the stopper sealed tight. What did you observe?

3) What did you see that indicated a chemical reaction took place?

4) Measure the mass of the flask and its contents. What is it?

5) How did this reaction show the conservation of mass?

6) Why did we seal the flask?

Conservation of Life: Photosynthesis and Respiration

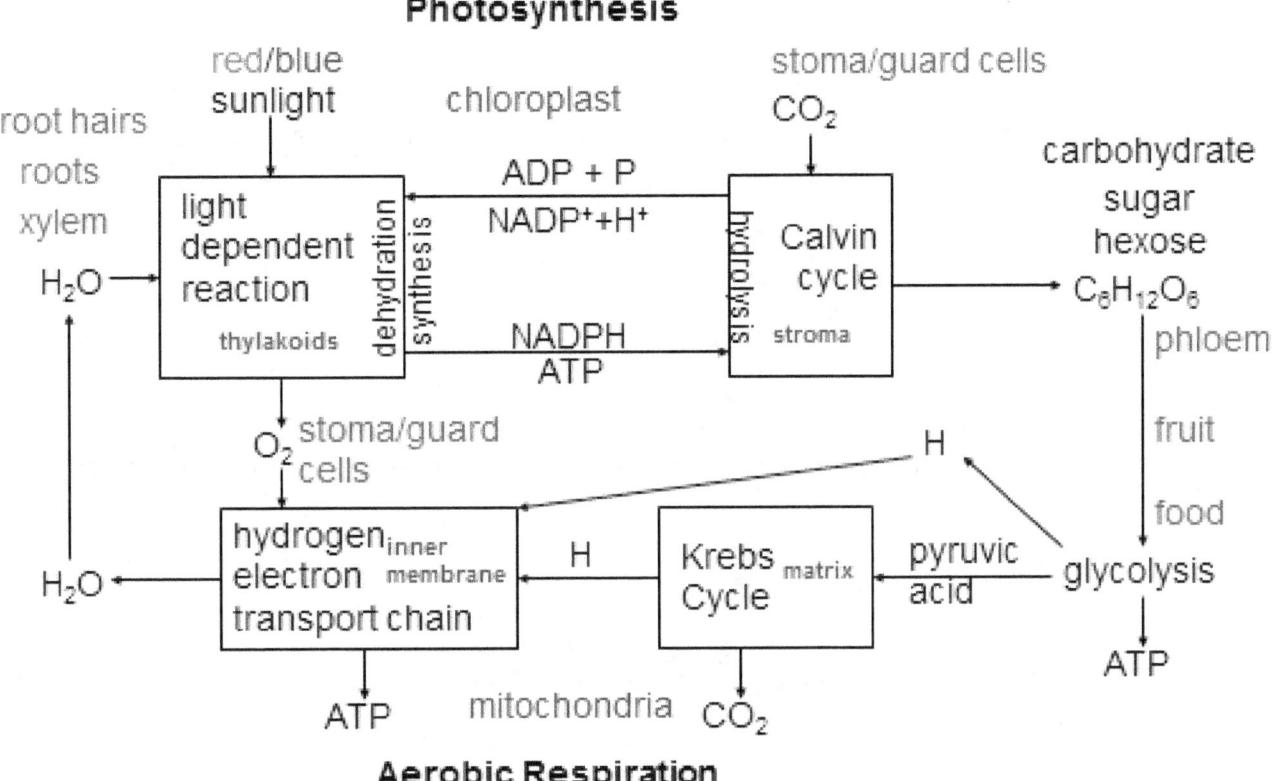

Directions:

Use the diagram above to answer the following questions.

1) Write the equation for photosynthesis and balance the reaction.

2) Write the equation for aerobic respiration and balance the reaction.

3) In both equations, trace where each element of the reactants go to make the products.

4) How are the two reactions similar to each other?

5) How does the conservation of mass, in this case, show that life has balance?

6) Plants and algae go through both photosynthesis and respiration. Animals only go through respiration. What would happen to life on Earth if we lost the plants and algae?

7) Keeping this in mind, which do you think formed first: the process of aerobic respiration or photosynthesis? Explain why.

8) How was this diagram a good model showing the conservation of mass in photosynthesis and respiration?

9) What careers would use the information shown by this activity?

Virtual Investigations that go with Chemical Reactions

ExploreLearning.com

    Chemical Changes

    Ionic Bonds

    Covalent Bonds

    Element Builder

    Electron Configuration

    Polarity and Intermolecular Forces

    Collision Theory

    Properties of Matter STEM Case

    Properties of Matter Handbook

    Physical and Chemical Changes STEM Case

    Physical and Chemical Changes Handbook

    Electrons and Chemical Reactions STEM Case

    Electrons and Chemical Reactions Handbook

    Stoichiometry

    Balancing Chemical Equations

    Limiting Reactants

    Moles

    Cell Energy Cycle

    Photosynthesis Lab

    Plants and Snails

Phet.colorado.edu

    Build an Atom

Build a Molecule

Molecule Polarity

Molecule Shapes

Molecule Shapes: Basics

Balancing Chemical Equations

Ratio and Proportion

Reactants Products and Leftovers

Physicsclassroom.com/Concept-Builders/Chemistry:

Classification of Matter

Name that Element

Particles... Words... Formulas

Formulas and Atom Counting

Chemical Equations

Atomic Models

Subatomic Particles

Isotopes

Periodic Table Battleship

Periodic Trends

Ionic Bonding

Bond Polarity

Molecular Polarity

Collision Model of Reaction Rates

Stoichiometry Relationships

Mole Conversions

Unit 5: Acids & Bases

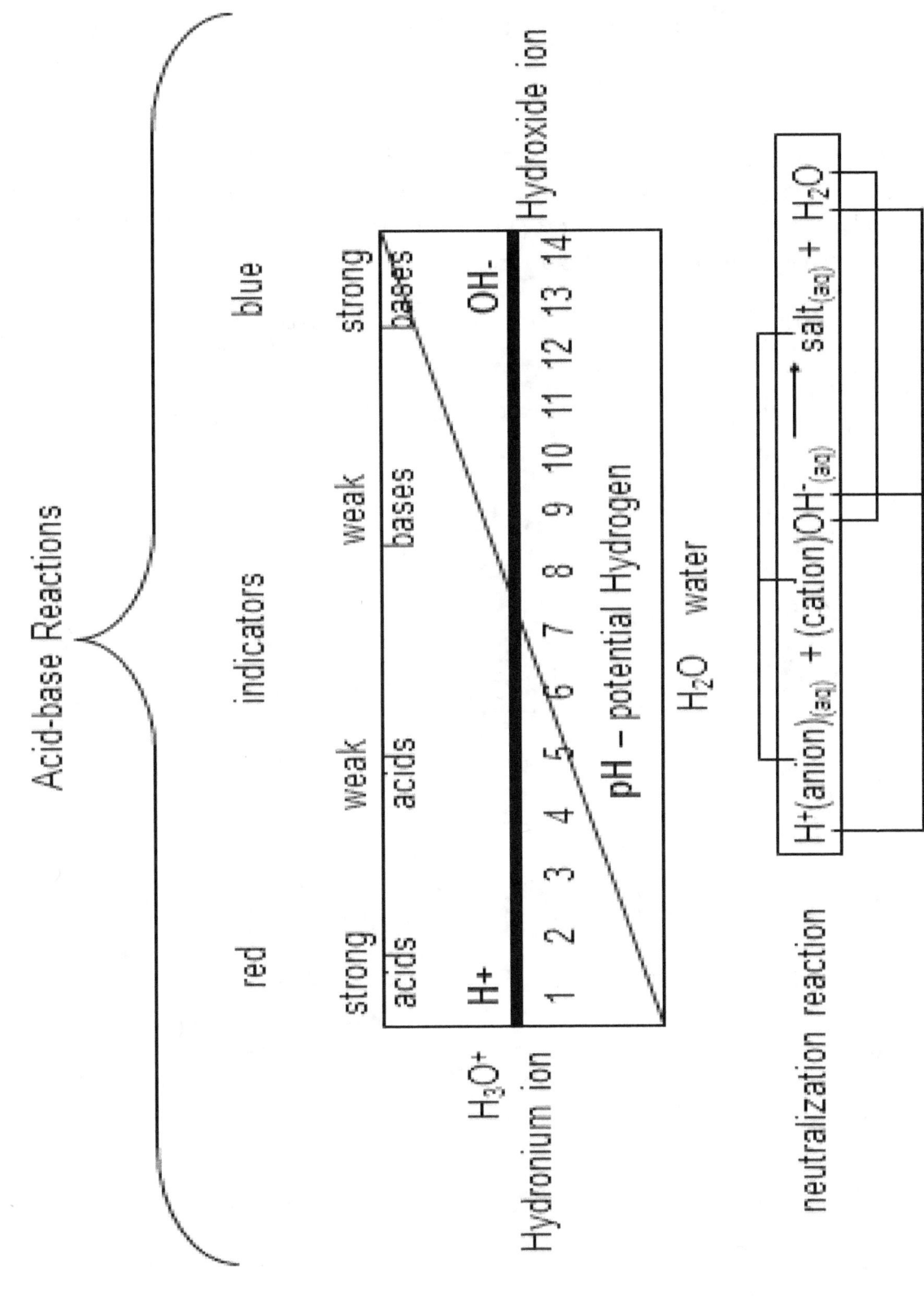

Which is an Acid and Which is a Base?

Directions:

Acids have H+ ions, and bases have OH- ions. Look at the formulas of some common acids and bases and see if you can tell them apart. Notice positive ions are written on the left side of the formula, and negative ions are written on the right side of the formula. Next to each formula, write down whether you think it is an acid or a base.

1) $Ca(OH)_2$

2) H_2SO_4

3) HCl

4) $NaOH$

5) $HC_2H_3O_2$

6) NH_4OH

7) KOH

8) H_2CO_3

9) $Ba(OH)_2$

10) $CsOH$

11) HNO_3

12) HBr

13) $Mg(OH)_2$

14) HI

15) $LiOH$

16) $Al(OH)_3$

17) $RbOH$

18) $HClO_4$

19) $HClO_3$

20) $Fe(OH)_2$

21) HBO_3

22) H_2S

23) $Zn(OH)_2$

24) HNO_2

25) $Fe(OH)_3$

26) HPO_3

27) $HC_2H_3O_2$

28) H_2SiO_3

29) How did you distinguish between an acid and a base with these compounds?

A Homemade Indicator

Directions:

You will need **safety goggles**, an **apron**, a **500 mL beaker**, five **small beakers**, **water**, **red cabbage**, a **hotplate**, **shampoo**, **grapefruit juice**, **Sprite**, **milk**, and **ammonia**. Looking at the materials and lab we will be using, what are the safety precautions we should take to protect ourselves and materials during the investigation?

1) Place about 300 mL of water in the 500 mL beaker with some leaves of purple cabbage. Boil it until the water turns purple.
2) In 5 small beakers, separately place shampoo, grapefruit juice, Sprite, milk, and ammonia.
3) The red cabbage juice is red in most acids and blue-purple in most bases. Use a pipet to mix the red cabbage juice with each substance to determine whether it is **acid** or a **base**.

Results:

4) What is shampoo?

5) What is grapefruit juice?

6) What is Sprite?

7) What is milk?

8) What is ammonia?

Observing Acid Relief

Directions and Questions:

You will need **100% purple grape juice**, a **beaker**, **water**, and an **antacid tablet**. Looking at the materials and lab we will be using, what are the safety precautions we should take to protect ourselves and materials during this investigation?

1) Take a beaker and fill it halfway up with a solution of water and purple grape juice. The pigment in grape juice is a natural indicator like red cabbage juice. It turns red in acids and a grayish blue in a base. What color is the grape juice in the water solution?

2) Is the solution an acid or base?

3) When stomach acid gets too strong, we take antacid tablets to calm our stomach down. Drop an antacid tablet into the water and grape juice solution. What do you notice happening?

4) Is the ending solution an acid or base?

5) How does an antacid tablet help calm our stomach or acid reflux?

6) Which careers might be interested in this type of reaction?

Acid or Base Grape Juice Indicator

Directions:

You will need **safety goggles**, an **apron**, a **pipette**, **100% purple grape juice**, seven tiny **beakers**, **vinegar, ammonia, lemon juice, Sprite, drain cleaner, detergent**, a **baking soda solution**, **litmus blue paper, litmus red paper, Universal indicator pH paper**, and a **pH meter** attached to an **interface** connected to a **computer** with **Logger Pro**. Looking at the materials and lab we will be using, what are the safety precautions we should take to protect ourselves and materials during the investigation?

1) Fill each of the seven beakers with one of these substances: vinegar, ammonia, lemon juice, Sprite, drain cleaner, detergent, and baking soda.
2) Use the red and blue litmus papers, dip them into each beaker, and write down what color they turn in Data Table 1.
3) Use the universal indicator pH paper, dip them into each beaker, and write down what pH the color indicated in Data Table 1.
4) Gently stir the pH meter in each beaker to see what the Logger Pro indicates is the pH of each solution. Be patient; it takes some time for the pH meter to stabilize. Make sure to rinse the pH meter between each beaker measurement. Write the pH measurement in Data Table 1.
5) Now take a pipette full of purple grape juice, squirt it into each beaker, and write down the color it makes in each of the solutions in Data Table 1.

Data Table 1

Solution	Blue Litmus	Red Litmus	Universal Indicator Paper	pH meter	Purple Grape Juice	Acid or Base?
Vinegar						
Ammonia						
Lemon juice						
Sprite						
Drain cleaner						
Detergent						
Baking soda						

Questions:

1) A pH below 7 indicates an acid; a pH above 7 indicates a base. What color does litmus paper turn when a solution is an acid?

2) What color does litmus paper turn when a solution is a base?

3) Determine if each substance is an **acid** or a **base** by filling in the last column in Data Table 1.

4) Which of the substances was the strongest acid (had the lowest pH)?

5) Which of the substances was the strongest base (had the highest pH)?

6) What color did the purple grape juice turn in the solution if it was an acid?

7) What color did the purple grape juice turn the solutions if it was a base?

8) Which of the indicators we used was most like the grape juice indicator? Explain.

9) Which of the indicators told you the most information? Explain.

10) Why do you think the purple grape juice is red when it is diluted?

Characteristics of Acids and Bases

Directions:

You will need **bottled water**, a **sink**, and put **vinegar, club soda, lemon juice, baking soda, soft soap,** and **laundry detergent** into **tiny cups** or **beakers**. Looking at the materials and lab we will be using, what are the safety precautions we should take to protect ourselves and materials during the investigation?

1) **Acids** are **sour,** and **bases** are **bitter** to the taste. **Bases feel slippery,** and **acids do not**. For each cup, separately dip your finger in and rub it on your thumb to find the feel (slippery or not slippery), and then lightly lick your finger to get a taste (bitter or sour). You can use the bottled water to wash your mouth out to get rid of the taste and use the sink to wash your fingers.
2) Circle the results in Data Table 1 below.

Data Table 1

Substance	Feel	Taste
Vinegar	Slippery or Not Slippery	Bitter or Sour
Club Soda	Slippery or Not Slippery	Bitter or Sour
Lemon Juice	Slippery or Not Slippery	Bitter or Sour
Baking Soda	Slippery or Not Slippery	Bitter or Sour
Soft Soap	Slippery or Not Slippery	Bitter or Sour
Laundry Detergent	Slippery or Not Slippery	Bitter or Sour

Questions:

1) Which substances were acids?

2) Which substances were bases?

Which will Corrode a Nail?

Directions:

You will need **two nails**, a small **bottle of coke** (you could also try **orange juice**), and a small bottle of **clear liquid soap**. **Looking at the materials and lab we will be using, what are the safety precautions we should take to protect ourselves and materials during the investigation?**

1) Acids tend to corrode metal, and bases do not. Coke has multiple acids in it, and soap is a base. **Hypothesis:** Which do you think will corrode the nail?

2) Open the lids of both the coke and the soap, place a nail in each one, and tightly fix the lids back on each.
3) Check them each day for two weeks to look for any signs of corrosion.
4) Fill in Data Table 1 below.

Data Table 1

Days	Corrosion of Nail in Coke	Corrosion of Nail in Soap
1	Yes or No	Yes or No
2	Yes or No	Yes or No
3	Yes or No	Yes or No
4	Yes or No	Yes or No
5	Yes or No	Yes or No
6	Yes or No	Yes or No
7	Yes or No	Yes or No
8	Yes or No	Yes or No
9	Yes or No	Yes or No
10	Yes or No	Yes or No
11	Yes or No	Yes or No
12	Yes or No	Yes or No
13	Yes or No	Yes or No
14	Yes or No	Yes or No

Question:

1) Did either show signs of corrosion in two weeks? If so, how?

2) Why do you think the pH of tap water is kept just over 7?

3) Why do you think soaps are good for cleaning metals?

4) Lemons contain lots of citric acid. Do you think lemon juice would corrode a nail? Explain.

Virtual Investigations that go with Acids and Bases

ExploreLearning.com:

    pH Analysis Gizmo

    pH Analysis: Quad Color Indicator Gizmo

    Titration Gizmo

Phet.colorado.edu:

    Acid-Base Solutions

    pH Scale

    pH Scale Basics

physicsclassroom.com/Concept-Builders/Chemistry:

    Which One Doesn't Belong? Acid-Base Properties

    Bronsted-Lowry Model of Acids and Bases

    pH and pOH

    Dissociation

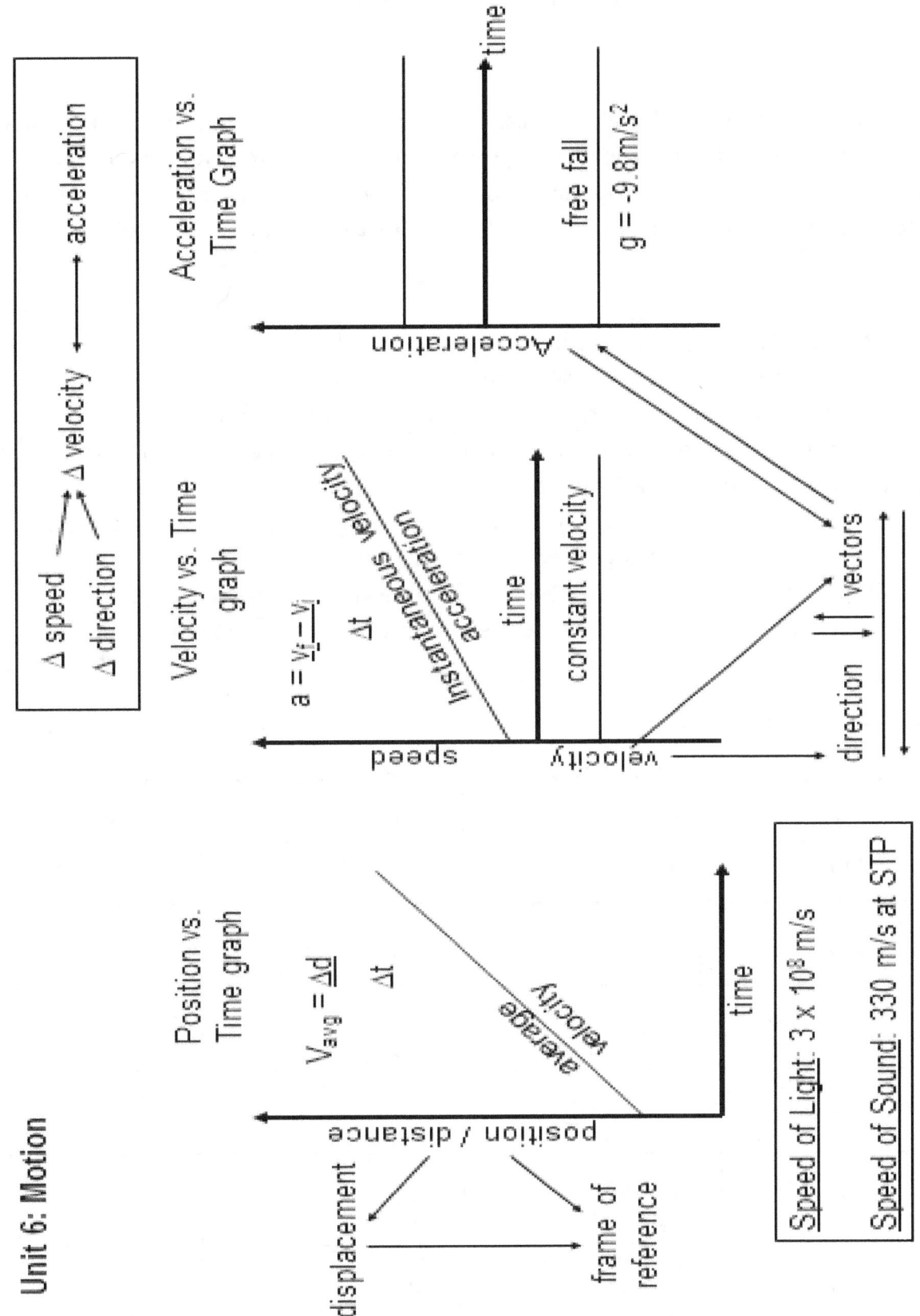

Simple Middle School Physical Science Investigations   Seven Sides Publishing

The Motion of a Bowling Ball

Directions:

You will need a **bowling ball** (an **air puck** can also be used), five **stopwatches**, a **large pillow** or something to act as a bumper, and **masking tape** to mark each meter with a **meter stick** on the floor for 5 meters. **Looking at the materials and lab we will be using, what are the safety precautions we should take to protect ourselves and materials during the investigation?**

1) Have five students line up one meter apart, each with a stopwatch.
2) At the far end, set up the bumper to stop the bowling ball.
3) Everyone starts the stopwatch when the bowling ball moves past the starting line.
4) When the ball passes each person at their meter line, they need to stop their stopwatch.
5) Record the times in Data Table 1 below.
6) Clear the stopwatches and repeat the procedure in # 3-5. Then calculate the averages for each distance by adding the two times and dividing by two.
7) Graph the averaged data in the Time vs. Distance graph on the next page.
8) Find the graph's slope by taking the rise (distance) divided by the run (time).

Data Table 1

Distance (m)	Trial 1 Time (s)	Trial 2 Time (s)	Average Time (s)
0 m	0 s	0 s	0 s
1 m			
2 m			
3 m			
4 m			
5 m			

Graph 1

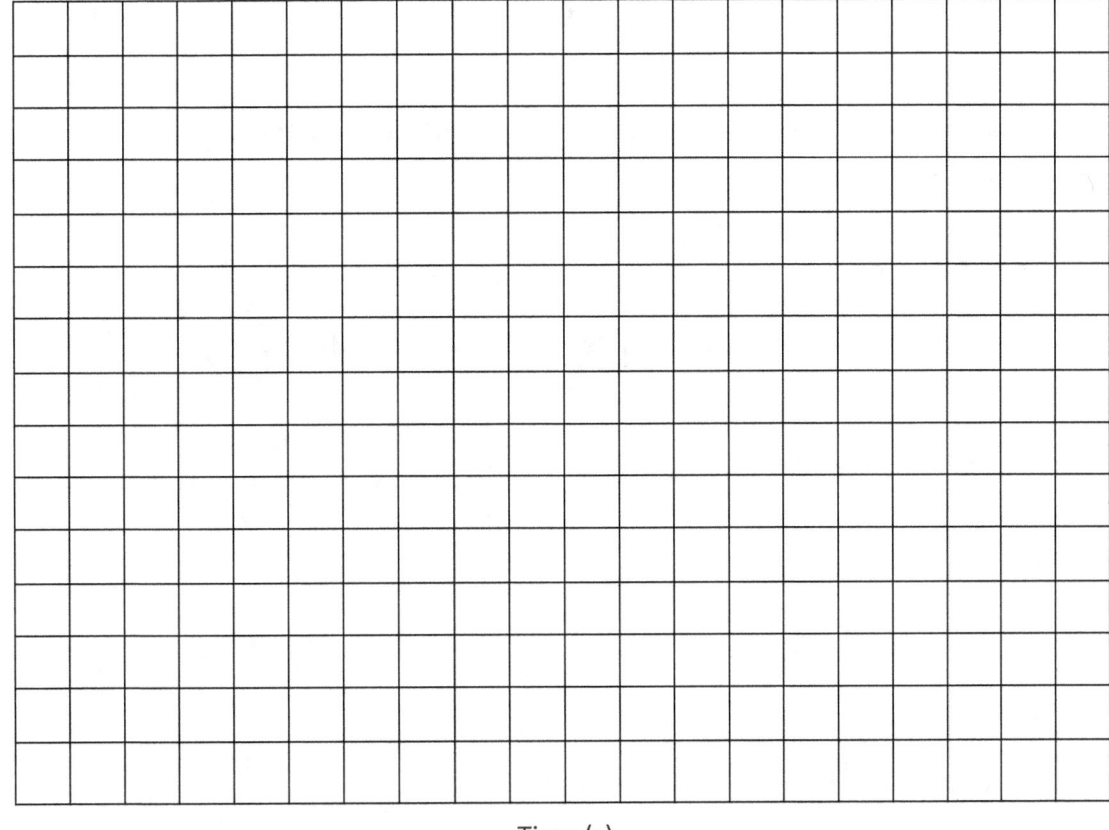

Questions:

1) On the distance-time graph, what does the slope of the line graph tell you?

2) What do you think a flat/horizontal line will tell you on a distance-time graph?

3) Imagine a bowling ball dropped from a great height. How would the motion of this ball relate to the one in the lab?

4) What was the average speed of the bowling ball?

5) Did the speed of the ball seem constant or changing?

6) Does this show the forces acting on the ball are balanced or unbalanced?

7) How far did the bowling ball travel while it was being timed?

8) What was its displacement?

9) How are distance and displacement related?

Marbles in Motion

Directions:

Get a segment of **hot wheels track, small stickers**, a **stopwatch**, and a **marble**. Have something that one side of the track can be placed on to raise that end to create a ramp. **Looking at the materials and lab we will be using, what are the safety precautions we should take to protect ourselves and materials during the investigation?**

1) Set the ramp so the ramp's bottom is along the edge of a tile on the floor. Most tiles in schools are 1 foot in length. Clear a path for 5 feet.
2) Adjust the height of the ramp so that the marble will just make it past 5 feet.
3) Place small stickers on the floor at the ramp's base and each foot past the ramp's base. The last one is 5 feet away from the base of the ramp.
4) Place a small sticker on the ramp to mark where you will place your marble for each trial to let it roll down the ramp; this keeps the distance your marble will be accelerating down the ramp constant.
5) Place your marble on the ramp and let it roll down (do not push). Time with a **stopwatch** how long it takes for the marble to move from the ramp's base to one foot away.
6) Repeat #5 four more times. Record the data in Data Table 1 below.
7) Repeat #s 5 and 6 for the distances 2 feet, 3 feet, 4 feet, and 5 feet away.
8) Find the average time for each distance.
9) Then calculate the average velocity for each distance by taking the distance and dividing it by the average time and write that in Data Table 1.

Data Table 1

Trial	1 foot	2 feet	3 feet	4 feet	5 feet
1					
2					
3					
4					
5					
Average Time					
Average Speed					

10) Take the average speed and plot them on the graph to make a speed-distance graph; this will look similar to a velocity-time graph since the longer the distance, the more time it takes. The shape we see will be the same as looking at accelerated motion on a velocity-time graph.
   a. This graph will be the same shape we would see for constant velocity motion on a distance-time graph.

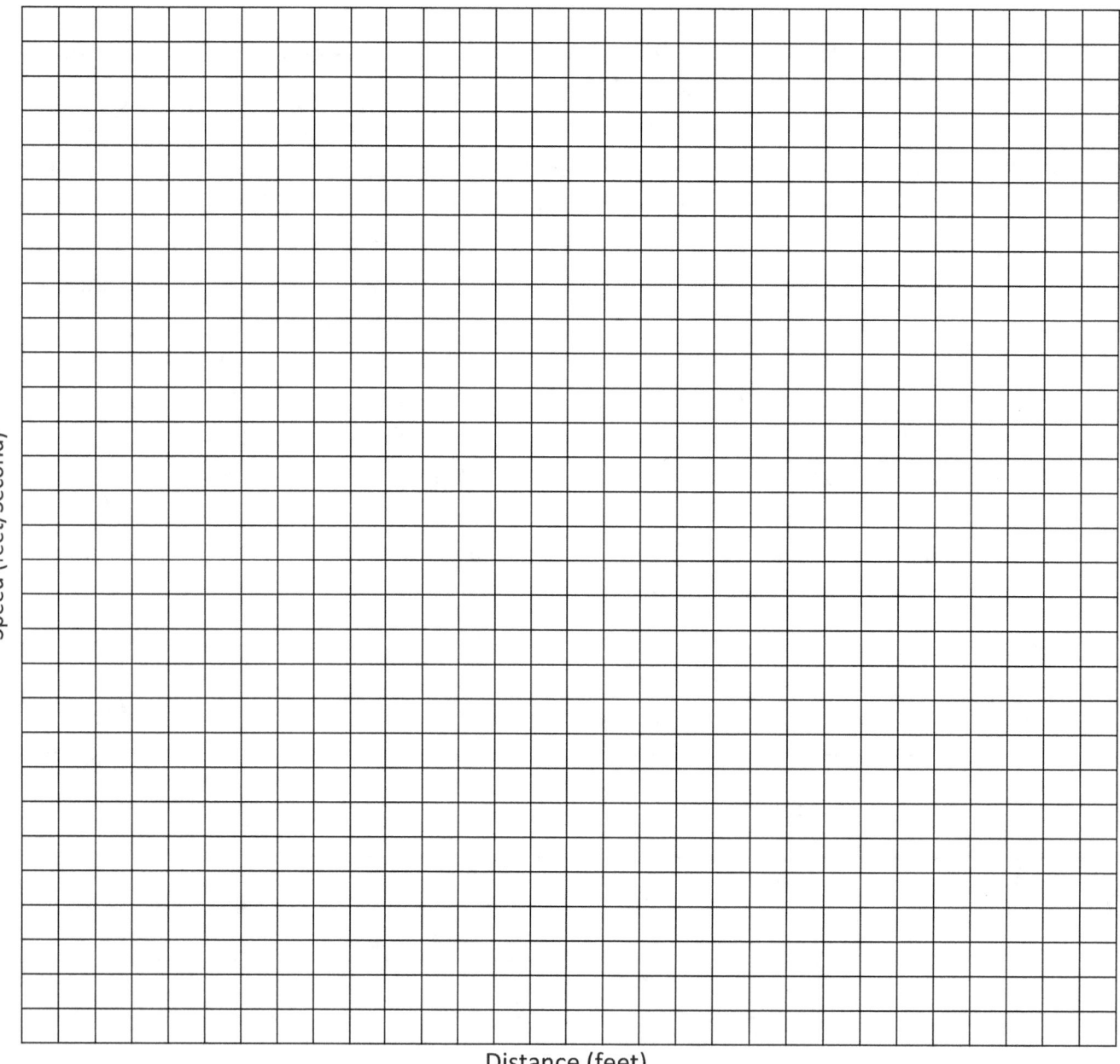

Velocity vs. Distance Graph

Questions:

1) Describe the motion of the marble as it moved down the ramp. If you have to, place the marble on the ramp and let it go to observe it move down the ramp.

2) What force caused the marble to speed up on the ramp?

3) How could we make the marble have a higher velocity at the bottom of the ramp?

4) Describe the motion of the marble as it moved across the floor. If you have to, place the marble on the ramp again and observe it roll across the floor.

5) What caused the marble to slow down across the floor?

6) How could we make the marble slow down faster across the floor?

7) At what distance would the average velocity be happening? At this distance would be the instantaneous velocity that is the same as the average velocity.

8) What was the shape of this graph? This shape of the graph is what acceleration looks like on a velocity-time graph.

9) What conditions would you need for the marble to have no positive or negative acceleration?

Ball Bounce

Directions:

You will need to get a **small wire basket,** a **large bouncy ball,** and attach a **motion detector** to an **interface** connected to a **computer** with **Logger Pro. Looking at the materials and lab we will be using, what are the safety precautions we should take to protect ourselves and materials during the investigation?**

1) Take the small wire basket and place it over the motion detector, so the ball (large bouncy ball) that you drop will not hit it. The spacing of the wires on the basket needs to be wide enough not to be detected by the motion detector.

2) When placing the basket over the motion detector, make sure the basket's gap is directly over the sensor, so the motion detector will only see the ball.

3) In Logger Pro, open folder Physics with Vernier and file #06 Ball Toss.

4) Press "Collect," drop the ball, and let it bounce on the basket. It is best if the ball bounces a few times to see what is happening in the data.

5) Look at the graphs and have the students see where each bounce is. Notice for each bounce; the graph sizes get less and less. Why would that happen?

6) Move the display to see only one bounce for all three graphs. Use that display to label the ball's motion on all three graphs simultaneously. Students can use the picture on the next page (similar to what you should see) to label what happened in the graphs they made.

7) Place a label on the graph where the ball is at the highest point, where the ball moves up, where the ball moves down, and where the ball is in contact with the basket.

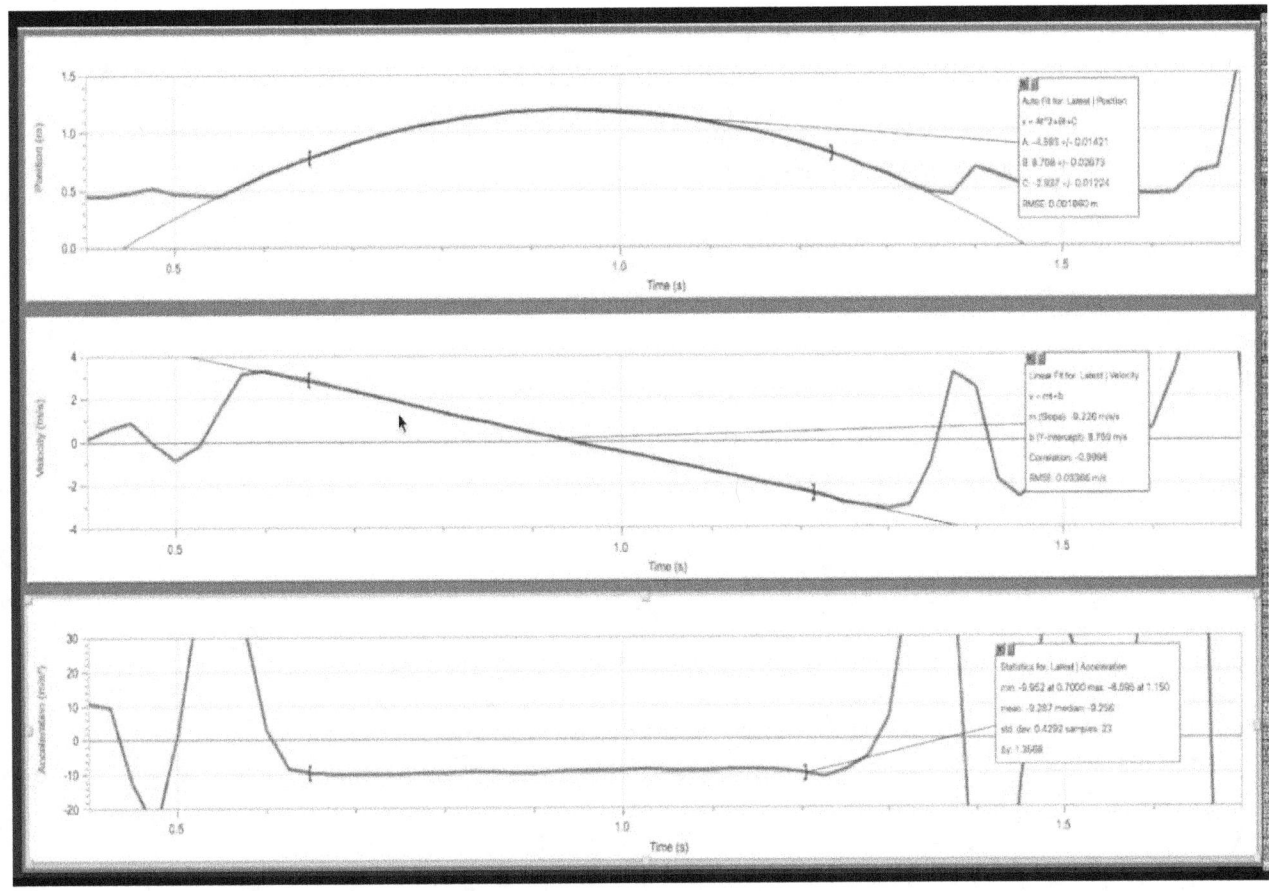

Questions:

1) What type of motion is this?

2) What is the shape of the graph of the ball as it goes through negative acceleration for:
   a. Distance vs. Time graph

   b. Velocity vs. Time graph

   c. Acceleration vs. Time graph

3) Compare this with the shape of constant velocity for:
   a. Distance vs. Time graph

b. Velocity vs. Time graph

c. Acceleration vs. Time graph

Cart on a Ramp

Directions:

You will need to get a **spring cart** with **Vernier's Dynamics System** with a **wall** on the end and a **motion detector** attached to an **interface** connected to a **computer** with **Logger Pro. Looking at the materials and lab we will be using, what are the safety precautions we should take to protect ourselves and materials during the investigation?**

1) Set up the spring cart on a ramp like one used in Vernier's Dynamics System, where one end of the ramp is propped up in the air.

2) Place the motion detector on the top end of the ramp with the sensor facing down the ramp.

3) In Logger Pro, open the folder Physics with Vernier and file #03 Cart on a Ramp.

4) Hold the cart at the top of the ramp with the spring side down and press "Collect," and let go of the cart, allowing it to roll down the ramp and bounce off the wall. It is best if the cart bounces a few times to see what is happening in the data.

5) Look at the graphs and have the students see where each bounce is. Notice for each bounce; the graph sizes get less and less. Why would that happen?

6) Move the display to see only one bounce for all three graphs. Use that display to label the cart's motion on all three graphs simultaneously. Students can use the picture on the next page (similar to what you should see) to label what happened in the graphs they made.

7) Place a label on the graph where the cart is at the highest point, where the cart moves up, where the cart moves down, and where the cart is in contact with the ramp's bottom wall.

Lab3Part1

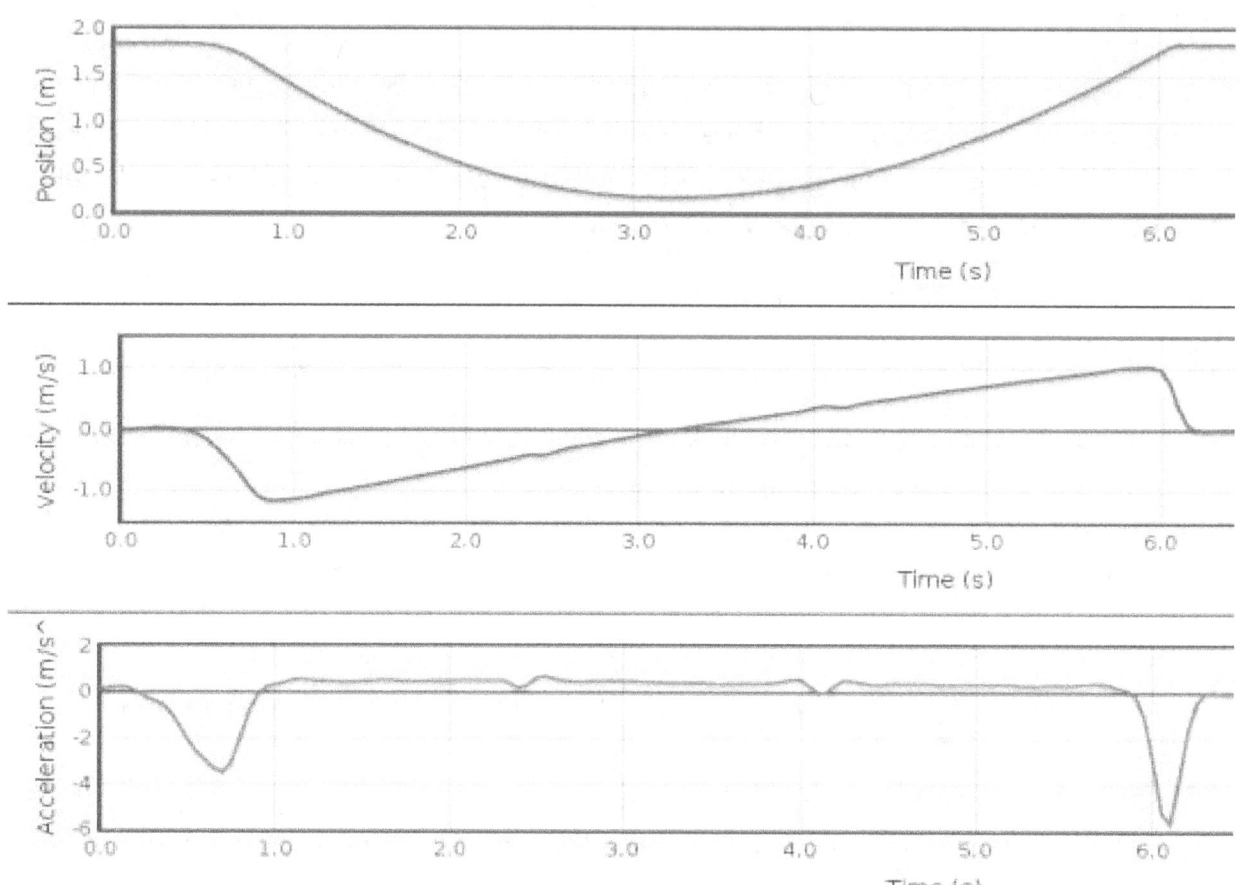

Questions:

1) What type of motion is this?

2) What is the shape of the graph of the cart as it goes through positive acceleration for:
   a. Position vs. Time graph

   b. Velocity vs. Time graph

   c. Acceleration vs. Time graph

3) Compare this with the shape of constant velocity for:
    d. Distance vs. Time graph

    e. Velocity vs. Time graph

    f. Acceleration vs. Time graph

Picket Fence Free Fall

Directions:

You will need Vernier's **Picket Fence**, a **photogate** attached to an **interface** connected to a **computer** with **Logger Pro**, and a **clamp** and **ring stand** to secure the photogate. **Looking at the materials and lab we will be using, what are the safety precautions we should take to protect ourselves and materials during the investigation?**

1) Use the clamp to fix the photogate to the ring stand and move the ring stand to the table's edge.

2) In Logger Pro, open the Physics with Vernier folder and file # 05 Picket Fence.

3) Click "Collect" to allow the photogate to turn on when you drop your picket fence through the photogate.

4) Then hold your picket fence over the photogate letting it hang down. Drop the photogate so that the alternating black and clear bands flow through the photogate. If one part of the picket fence does not go through in the drop, your reading will be off.

5) Look at your data, and make sure your slope for the velocity vs. time graph is a straight line; if it isn't, repeat #s 3 & 4.

6) If the graph's slope is straight, click the linear fit button and record the line's slope in the top data table on the next page; this is the acceleration of the picket fence dropping through the photogate.

7) Repeat steps 3-6 four more times.

8) Find the Minimum and Maximum values of the trials, calculate the average by adding up all five slopes, and then dividing by 5. Write this value in the middle data table on the next page. It should be close to 9.81 m/s².

9) Find the precision by taking the lowest number of your average and 9.81 and dividing by the other, then multiply by 100, giving you the % accuracy. Write that down in the bottom data table.

Simple Middle School Physical Science Investigations

Data Tables

Trial	1	2	3	4	5
Slope (m/s²)					

	Maximum	Minimum	Average
Acceleration (m/s²)			

Acceleration due to gravity, g	9.81 m/s²
Precision	%

Questions:

1) How does the force of gravity affect the motion of a falling object?

2) How does this relate to Newton's 2nd Law of Motion?

3) What is the shape of the position vs. time graph for each trial?

4) What is the shape of the velocity-time graph for each trial?

5) How close is your average acceleration compared to the 9.81 m/s²?

Simple Middle School Physical Science Investigations   Seven Sides Publishing

Elevator Lab

Directions and Questions:

Take a **1 kg mass** into an **elevator** with a **digital scale. Looking at the materials and lab we will be using, what are the safety precautions we should be taking to protect ourselves and materials during the investigation?**

1) Place the 1 kg mass on the digital scale on the elevator floor.
2) What is the mass in the scale say before the Elevator moves?

3) Press a button in the elevator to move the elevator down. How does the reading on the scale change when the elevator starts to move down?

   (gets higher, gets lower)

4) How does the reading on the scale change when the elevator starts to slow down?

   (gets higher, gets lower)

5) Press a button in the elevator to move the elevator up. How does the reading on the scale change when the elevator starts to move up?

   (gets higher, gets lower)

6) How does the reading on the scale change when the elevator starts to slow down?

   (gets higher, gets lower)

7) When did the mass seem to have less weight?

8) Try to explain why.

9) When did the mass seem to have more weight?

10) Try to explain why.

11) How did this investigation show Newton's 2nd Law of Motion?

12) Where in this investigation did we see Newton's 1st Law of Motion?

13) Where in this investigation did we see Newton's 3rd Law of Motion?

Virtual Investigations that go with Motion

ExploreLearning.com:

    Measuring Motion Gizmo

    Distance-Time Graphs Gizmo

    Distance-Time and Velocity-Time Graphs Gizmo

    Vectors Gizmo

    Adding Vectors Gizmo

    Pythagorean Theorem Gizmo

    Pythagorean Theorem with Geoboard Gizmo

    Cat and Mouse (Modeling with Linear Systems) Gizmo

    Distance Formula Gizmo

    Free Fall Tower Gizmo

    Sled Wars Gizmo

    Atwood Machine Gizmo

    Free-Fall Laboratory Gizmo

    Feed the Monkey Gizmo

    Golf Range Gizmo

PhET.colorado.edu:

    Vector Addition

    Maze Game

    Motion 2D

    The Moving Man

    Ladybug Motion

    Projectile Motion

Physicsclassroom.com:

    Physics Interactives:

        Vector Walk

        Vector Addition

        Name that Vector

        Vector Guessing Game

        Vector Addition: Does Order Matter?

        Riverboat Simulation

        Kinematics

        Graphs & Ramps

        Vectors and Projectiles

        Projectile Simulator

        Turd the Target

        Turd the Target 2

        Monkey and the Zookeeper

    Concept Builders:

        Kinematics

        Distance vs. Displacement

        Speed-Distance-Time

        Motion Diagrams

        Position – Time Graphs – Conceptual Analysis

        Position – Time Graphs – Numerical Analysis

        Acceleration

        Name That Motion

Graph That Motion

Match That Graph

Velocity – Time Graphs

Dots and Graphs

Words and Graphs

Which One Doesn't Belong

Vectors and Projectiles

Vector Direction

Head–To–Tail Vector Addition

Vector Addition

Component Addition

Free Fall

Up and Down

Which One Doesn't Belong? Projectile Motion

Trajectory – Horizontally Launched Projectiles

Trajectory – Angle Launched Projectiles

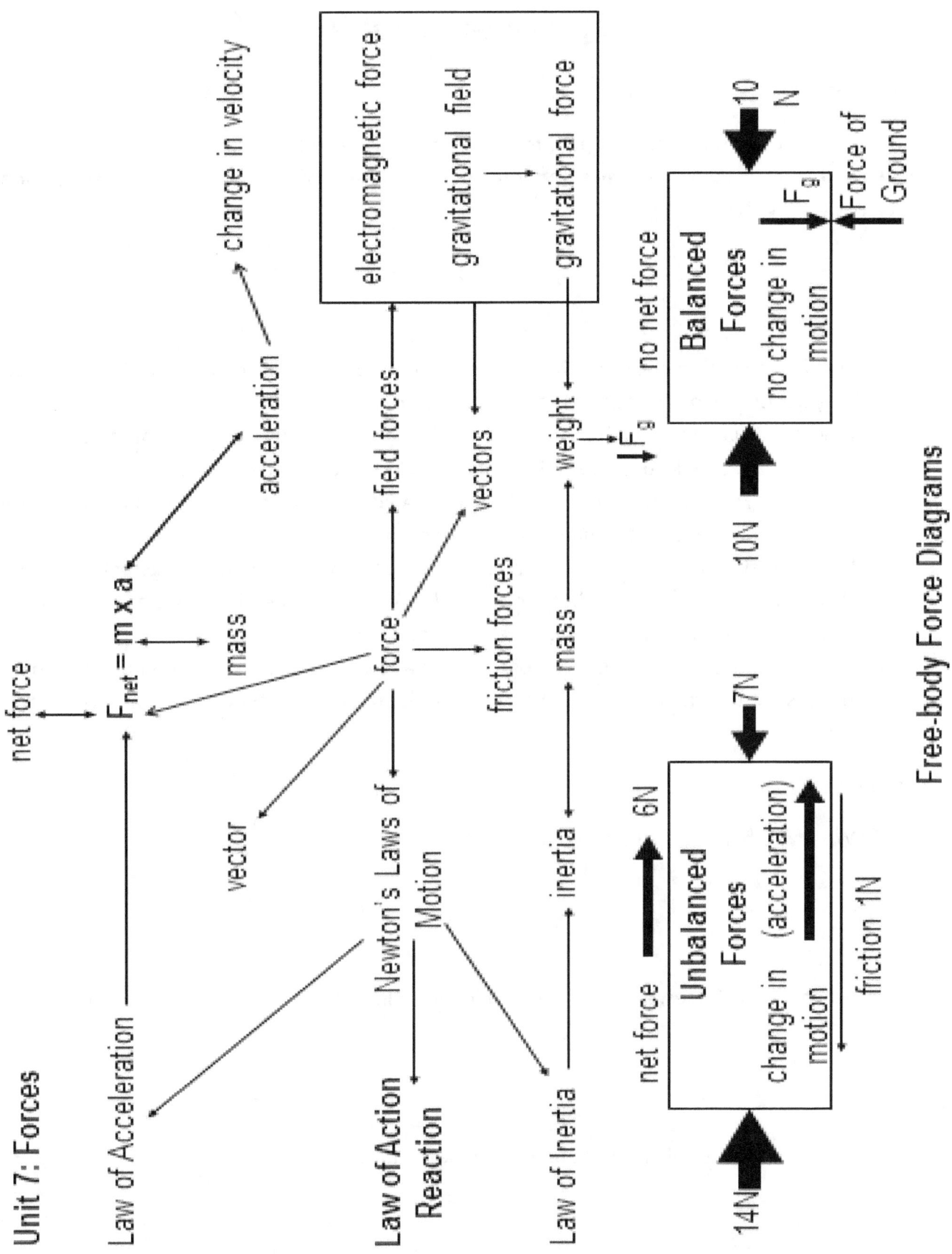

The Human Table

Directions:

You will need four **chairs** and four **people** about the same height. **Looking at the materials and lab we will be using, what are the safety precautions we should take to protect ourselves and materials during the investigation?**

1) Set the four chairs facing each other in a square a couple of feet apart. Distances can vary depending on the size of the people you are using.
2) Have the students sit on the chairs with their legs on the right-hand side of the chairs.
3) The student should then be able to lean back and place their heads on the legs just above the knees of the person behind them. Adjust the chairs if you need to make sure each will be able to do this before they lean back.
4) Have the students lean back and place their heads on the person's legs behind them. Have each person lift their butts off the chairs and pull the chairs out away from the student square. You should now have a human table.
5) The students should be able to hold this for a few minutes, discuss the questions with the rest of the class, then put the chairs back under the butts of the students so they can sit up.

Questions:

1) What was holding the students up?

2) What are the forces, and how are they involved in this system?

3) Were the forces balanced or unbalanced? Explain why.

Balancing Forks

Directions:

You will need two **forks**, a **toothpick**, and a **glass**. **Looking at the materials we will be using, what safety precautions should we take to protect ourselves and materials during the investigation?**

1) Take the two forks and force them together at the prong end.
2) Take a toothpick and place it on the rim of the glass; simultaneously, balance the two forks on the toothpick with the forks' handles going around the sides of the glass.
3) Discuss with the class and teacher why you think this is happening.

Questions:

1) What do you think is causing the forks to be balanced on the toothpick without falling?

2) What forces are acting on this system?

   a. How are they involved in this system?

3) Draw a force diagram of this system below showing the forces and how they balance the system.

Comparing Friction Lab

Directions:

You will need an **ice cube, rock, eraser, wooden block, aluminum foil**, and a **tray. Looking at the materials and lab we will be using, what are the safety precautions we should take to protect ourselves and materials during the investigation?**

1) Position each object on your tray.
2) Slowly lift one end of the tray and stop when an object slides.
3) Measure the height of the end of the tray you raised.
4) Keep doing this until you have measurements for all of your objects.
5) Put all your data in Data Table 1 below.

Data Table 1

Object	Height at which the object slid (cm)
Ice Cube	
Rock	
Eraser	
Wooden block	
Aluminum foil	

Questions:

1) Why did the objects slide off at different heights?

2) What type of friction did each object overcome to start sliding?

3) What type of friction does the object have as it slides down the tray?

4) Draw a force diagram showing the motion and forces of the objects as they slide down a ramp.

5) Which direction is the friction force from the movement?

Measuring the Effects of Air Resistance

Directions:

You will need two **stopwatches** and two **pieces of paper. Looking at the materials and lab we will be using, what are the safety precautions we should take to protect ourselves and materials during the investigation?**

1) Take one piece of paper and wad it up into a ball. Leave the other paper flat.
2) Hold both pieces of paper at the same height and drop them simultaneously. Make sure the flat paper is horizontal to the floor when you drop it. Have one person time the flat paper and the other person time the paper wadded up in a ball.
3) Write the times in Data Table 1 below.
4) Repeat the procedure in #s 2-3 two more times.
5) Find the averages of the times by adding the three times up and dividing by 3. Put this data in Data Table 1 below.

Data Table 1

Type of Paper	Trial 1 Time (s)	Trial 2 Time (s)	Trial 3 Time (s)	Average Time (s)
Flat				
Wadded				

Questions:

1) Which paper accelerated faster?

2) Which one reached its terminal velocity first?

3) What were the forces acting on the papers?

4) How does this relate to why we use parachutes when jumping out of planes?

5) Which one accelerated faster?

6) Which one reached its terminal velocity first?

7) Which part of the investigation showed Newton's 1st Law of Motion?

8) Which part of the investigations showed Newton's 2nd Law of Motion?

9) Which part of the investigation showed Newton's 3rd Law of Motion?

10) How does this relate to why we use parachutes when jumping out of planes?

11) Which paper showed balanced forces? Explain why.

12) Which paper showed unbalanced forces? Explain why.

13) Draw a force diagram showing the magnitude of the forces acting on the flat paper once it reached terminal velocity.

14) Draw a force diagram showing the forces acting on the paper ball accelerating to the ground.

15) Which careers need to know this information to protect people and equipment?

Simple Middle School Physical Science Investigations   Seven Sides Publishing

Air Resistance

Directions:

You will need a **scale**, five **coffee filters**, a **ring stand**, and a **motion detector** attached to an **interface** connected to a **computer** with **Logger Pro**. **Looking at the materials and lab we will be using, what are the safety precautions we should take to protect ourselves and materials during the investigation?**

1) Place the motion detector on the ring of the ring stand facing down.
2) In Logger Pro, open the folder Physics with Vernier and file 13 Air Resistance.
3) Find the coffee filter(s) mass and place that in Data Table 1 below.
4) Hold a coffee filter between .25m - .5m away from the motion detector.
5) Click "Collect" to begin data collection.
6) Drop the coffee filter so it will fall directly down under the motion detector.
7) The slope of the straight line will be the terminal velocity. Select the straight slope line and click the linear fit button to tell you the slope. Write that number in the data table below for one filter.
8) Repeat steps 3 through 7, adding another coffee filter for each trial, and put those numbers in Data Table 1 below.

Data Table 1

Number of Filters	Mass of the Filters (g)	Terminal Velocity (m/s)
1		
2		
3		
4		
5		

Questions:

1) How does the mass of the object affect terminal velocity?

2) What forces are acting on the filters?

3) How does inertia Newton's 1st Law of Motion affect terminal velocity?

4) How could surface area affect terminal velocity?

5) Does everyone need the same size parachute?

6) How big of a parachute would an elephant need compared to a human (Operation Dumbo Drop)?

7) Draw a force diagram showing balanced forces as a parachute would slow the elephant to a terminal velocity as it drops to the ground

Observing Inertia, Newton's First Law of Motion

Directions:

You will need a **toy car that winds up** when you pull it back and then moves forward when you let it go. You will also need a **penny** and a **rubber band. Looking at the materials and lab we will be using, what are the safety precautions we should take to protect ourselves and materials during the investigation?**

1) Place the penny on the car. Pull the car back to wind up the car. Let it move forward and have it crash into something.

2) What happens to the penny?

3) Why did the penny do this?

4) Place a rubber band around the car and use it to fix the penny to the car.
5) Repeat the procedure in # 1
6) What happens to the penny?

7) Why did the penny do this?

Questions:

1) Explain the forces acting in both crashes.

2) What does the penny represent?

3) What does the rubber band represent?

4) Why should we wear seat belts when we get in a car?

Inertia Lab Stations

Equipment and Safety:

You will need two **ping pong balls**, a **ping pong paddle**, two **tennis balls**, an **index card**, a **cup**, and a **penny**. Looking at the materials and lab we will be using, what are the safety precautions we should take to protect ourselves and materials during the investigation?

Remember that **inertia** is the resistance of objects to change their motion. Objects in motion tend to stay in motion, and objects at rest tend to stay at rest; unless acted on by an outside force (Newton's 1st Law of Motion). The larger the mass, the more inertia it has. Think of this as you go through the three lab stations.

Station 1

Take a **ping pong ball** and use the **paddle** to bounce it off the door or wall a few times. Now take a **tennis ball** and use the paddle to bounce it off the door or wall a few times.

1) Which ball seems to have more inertia?

2) How can you tell?

3) Why does that ball have more inertia?

4) Which ball can you get to move faster off the paddle? Why?

Station 2

Place an **index card** on the opening of the **cup**. Now place a **penny** on top of the index card. With one hand, hold the cup; with the other hand, grab the index card and quickly pull it away horizontally.

1) What happened to the penny?

2) Why did it not move horizontally with the card?

3) What forces acted on the penny after the card was pulled away?

Station 3

1) Roll a **ping pong ball** at a stationary **tennis ball**, have them collide, and describe what happens.

2) Now roll the tennis ball at a stationary ping pong ball, have them collide, and describe what happens.

3) Use your knowledge of the Law of Inertia to explain why the results were not the same.

Observing Centripetal Force

Directions:

Put some **water** in a **bucket** with a strong handle and get a **penny** and a **wire coat hanger**. Looking at the materials and lab we will be using, what are the safety precautions we should take to protect ourselves and materials during the investigation?

1) Stand away from others. The best place to go is outside.
2) Swing the bucket in a circle makes the bucket go upside down at one point. Make sure it keeps moving during the observation.

Questions:

1) Why did you not get wet?

2) Which direction does the water want to go?

3) Why does it want to go there?

4) What force does your arm represent if you represented the sun and the bucket represented the Earth?

5) What would happen if you swung the bucket slowly?

6) Was the bucket moving at a constant velocity, or was it accelerating? Explain why.

7) How does this investigation show how and why the Earth orbits the Sun, and the Moon orbits the Earth?

8) Which planet in the solar system would be the hardest to take out of orbit because of inertia? Explain why.

9) Which planet would be the easiest to take out of orbit because of its inertia? Explain why.

10) Take the wire coat hanger and bend the triangle's bottom side to make a square so you can spin the hanger around your finger there. You may have to file down the end of the hook so you can balance a penny on it. Once the penny is balanced (works best tails side down), gently spin the hanger around your finger. If you do it right, the penny should stay balanced on the hanger. Explain why this happens.

Centripetal Force Under Glass

Directions:

You will need a **glass** or a **see-through plastic cup** and a **marble** on a tabletop. **Looking at the materials and lab we will be using, what are the safety precautions we should take to protect ourselves and materials during the investigation?**

1) Place the marble in the glass/plastic cup. Move the glass/cup in a circular motion causing the marble to spin around the inside of the glass/cup off the table's surface.
2) Discuss what you are seeing with the class and your teacher.

Questions:

1) Explain why the marble can stay off the surface of the table.

2) Where is the inertia in this lab?

3) What happens when you stop the movement of your glass/cup?

4) How does this lab show how inertia plays a role in planets orbiting the sun?

5) Was the marble moving at a constant velocity, or was it accelerating? Explain why.

Newton's Relay Race

Directions and Questions:

You will need a **broom**, a **bowling ball**, a **basketball**, and a kid's **rubber ball. Looking at the materials and lab we will be using, what are the safety precautions we should take to protect ourselves and materials during the investigation?**

Accelerating an object from rest: we will be observing <u>inertia</u> – resistance to change motion (Newton's 1st Law)

1) Place the bowling ball on the floor. Push the ball with a broom in a sweeping motion to cause the bowling ball to accelerate. When does the bowling ball accelerate?

2) How easy was it to accelerate?

3) When does it move at a constant speed?

4) Place the kid's rubber ball on the floor and push this ball the same way you did with the bowling ball. When does the ball accelerate?

5) How easy was it to accelerate?

6) When does the ball move at a constant speed?

7) Which ball had the most inertia?

8) Draw a force diagram of the action of accelerating the ball.

Stopping an object: observing inertia (Newton's 1st Law) and <u>momentum</u> – resistance to stop motion

9) Get the bowling ball moving, then stop it with the broom. Do the same with the kid's rubber ball. Which ball was harder to stop?

10) When sitting still, which object has the most inertia?

11) Draw a force diagram of the action of stopping a moving ball.

Turning an object 180 degrees: observing inertia (Newton's 1st Law) and momentum

12) Get the bowling ball and start moving it, then stop it and turn it 180 degrees. Do the same for the kid's rubber ball. Which ball was easier to change direction?

13) Why do you think that ball was easier to change direction?

14) Which ball had the most inertia?

15) Which ball accelerated the fastest?

Applying a constant force on the ball: observing force and acceleration (Newton's 2nd Law)

16) Go to a long hallway and make sure it is clear. Get the bowling ball moving by pushing it with the broom; see what happens if you try to put a constant force on the ball. Do the same for the kid's rubber ball. What happened to the speed of both balls?

17) Could you keep doing it?

18) Which ball could you apply the force for the longest amount of time? Why?

19) Which ball accelerated the fastest?

20) How could you make the balls move at a constant velocity?

Relay Race: observing inertia (Newton's 1st and 2nd Laws)

21) Make a course that changes direction several times (I have made my student go around the center demo table in my room) to push the different balls around in a race. Divide up into three equal teams, each with a broom. One team will push a bowling ball with a broom, one team will push a basketball with a broom, and one group will push the kid's rubber ball with the broom. Predict which team will finish the course first with all its team members.

22) Have the students do a relay race to move the balls through the course (be very careful of the bowling ball and make sure kids know to move out of its way if it comes at them). Which team won?

23) Why were they able to win?

24) Which team came in last? Why?

25) When was Newton's 3rd Law of Motion happening during today's investigation?

Newton's Second Law

Directions:

You will need a **scale**, a **cart**, a long **rubber band**, a **dual-range force sensor**, and **low g accelerometer**, and an **interface** connected to a **computer** with **Logger Pro**. **Looking at the materials and lab we will be using, what are the safety precautions we should take to protect ourselves and materials during the investigation?**

1) Stack the dual-range force sensor on top of the cart and the accelerometer on the force sensor. Have them face the same direction as the wheels will be moving, so the force and acceleration will be measured in the same direction. Tightly wrap the rubber band around the system, holding it all together.

2) Connect the dual-range force sensor to channel 1 on the interface. Connect the low g accelerometer to channel 2.

3) Open Physics with Vernier folder and file # 09 Newton's Second Law.

4) Click "Collect" to collect data. Holding the hook of the dual-range force sensor, roll the cart back and forth in the direction the wheels move. Vary the forces, both small and large.

5) What is the shape of the force vs. time graph?

6) What is the shape of the acceleration vs. time graph?

7) Click the examination button. Move the mouse across one of the graphs. When the force is at its maximum, what is the acceleration? (maximum or minimum)

8) Click on the Force vs. Acceleration graph and click the linear fit button. Record the slope of this line in Data Table 1.

9) Find the mass of the cart and sensors. Write this in Data Table 1.
10) Add/fix a 500 g mass to the cart, repeat the procedure for #4 and 8, and record the slope of the force vs. acceleration graph in Data Table 1.

11) Find the mass of everything in the cart now. Write that in Data Table 1.

Data Table 1

Cart	Slope of graph	Mass (kg)
Cart and Sensors		Kg
Cart, sensors, and 500 g		Kg

Questions:

1) How are the Force vs. Time and Acceleration vs. Time graphs similar for the two trials?

2) How are they different?

3) Compare the slope of the Force vs. Acceleration graph and the mass. What does the slope represent?

4) Write a general formula for the three variables: force, mass, and acceleration.

5) What is the unit for the slope of the Force vs. Acceleration graph?

6) What is the relationship between force and acceleration in the equation?

7) What is the relationship between the mass and acceleration in the equation?

8) How could this information help you run away from a rhinoceros chasing you?

9) If you have a bowling ball and a baseball, each is suspended by a separate rope and hit each with a baseball bat, which ball will have the biggest change in motion? Explain why.

Fan Cart Lab

Directions:

You will need a **scale**, a **mass**, a **cart**, a **fan** to fix to the cart, a **motion detector** attached to an **interface** connected to a **computer** with **Logger Pro**, and a **Dynamics System** from Vernier. **Looking at the materials and lab we will be using, what are the safety precautions we should take to protect ourselves and materials during the investigation?**

1) Use the scale to measure the mass of the cart. Write this in Data Table 1.
2) Position the motion detector at one end of the track. Position the cart in front of the motion detector so it will move away from the motion sensor.
3) Start the fan. Click the "Collect" button. Allow the cart to accelerate down the track. Click "Stop." Grab the cart before it falls off the track.
4) Highlight the line on the Velocity vs. time graph. Click the "Fit" button and choose linear fit. The slope of the line is the acceleration of the cart. Record the acceleration in Data Table 1.
5) Repeat #3-4 two more times and put the accelerations in Data Table 1.
6) Calculate the average acceleration by adding the three values and dividing by 3.
7) Add a mass to the fan cart. Record the amount of the mass in the data below.
8) Repeat the procedures in #3-6 with the added mass. Put this data in Data Table 1.

Data:

Mass of the cart and fan _____ kg   Amount of mass added _____ kg

Data Table 1

Cart	Mass (kg)	Trial 1 Acceleration	Trial 2 Acceleration	Trial 3 Acceleration	Average Acceleration
Cart + Fan					
Cart + Fan + Mass					

Questions:

1) Using the formula F = m x a, calculate the force of the fan on the cart. What is that force?

2) Do the same for the cart with the mass on it. What is that force?

3) Compare the answers for #s 1 and 2.

4) Why did they have different accelerations then?

5) What may be a source of error not figured in the equation?

Simple Middle School Physical Science Investigations   Seven Sides Publishing

Newton's Third Law

Directions and Questions:

You will need a **rubber band** and two **dual-range force sensors** attached to an **interface** connected to a **computer** with **Logger Pro**. **Looking at the materials and lab we will be using, what are the safety precautions we should take to protect ourselves and materials during the investigation?**

1) Hold a rubber band between your right hand and your left hand. Pull with your left hand. Does your right hand experience a force?

2) Does your right hand apply a force to the rubber band?

3) What direction is this force compared to the left hand?

4) Pull harder with your left hand. Does this change any force applied by your right hand?

5) How is your left hand's force, transmitted by the rubber band, related to the force applied by your right hand?

6) Write a rule in words for this force relationship.

7) Place the two dual-range force sensors opposite each other with their hooks facing each other. Attach the rubber band between them.

8) In Logger Pro, open the folder Physics with Vernier and the file 11 Newton's 3rd Law.

9) Press "Collect" and have one person pull on one sensor and another person pull on the other back and forth, stretching and shortening the rubber band. What does the graph show about the magnitude of both forces?

10) What does the graph show about the direction of the two forces?

11) Is there any way to pull on your force sensor without your partner's force sensor pulling back while keeping tension on the rubber band?

12) Fasten one force sensor to your lab table and repeat the experiment. Does the lab table pull back?

13) Connect the two force sensors together with just their hooks instead of the rubber band. How do the results change?

14) State Newton's Third Law of motion:

15) How does this lab show Newton's Third Law of Motion?

16) If you are driving down the street and a bug splatters on your windshield. What is greater: the force of the bug on the windshield or the force of the windshield on the bug? Explain why.

Water Bottle Rockets

Directions and Questions:

You will need a **water bottle rocket launcher**, a **2-liter bottle**, and an **air pump** with a **pressure gauge. Looking at the materials and lab we will be using, what are the safety precautions we should take to protect ourselves and materials during the investigation?**

1) Fill the water bottle half full with water.
2) Angle the launcher straight up at a 90-degree angle to the ground (gives the rocket its highest distance to travel in the air).
3) Pump 20 pounds of pressure into the rocket. Start a stopwatch when you launch the rocket and stop it when it reaches the highest point in the air. What was the time?

4) What do you think will happen to the time in the air and the launch's height if we double the pressure?

5) Pump 40 pounds of pressure into the rocket. Start a stopwatch when you launch the rocket and stop it when it reaches the highest point in the air. What is the time?

6) How did changing the pressure affect the flight of your rocket?

7) If you were to launch again, what other variables could we change to affect the rocket's height?

8) How do you think changing the rocket's mass will affect the force of gravity?

9) How would it affect the inertia?

10) How is Newton's 1st Law of Motion affect the rocket launch?

11) How does Newton's 2nd Law of Motion affect the rocket launch?

12) Draw a force diagram to show the variables affected in Newton's 2nd Law of Motion.

13) How is Newton's 3rd Law of Motion seen in the launch?

14) Draw a force diagram to show how the launch shows Newton's 3rd Law of Motion.

Virtual Investigations that go with Force

ExploreLearning.com

    Force and Fan Carts

    Fan Cart Physics

    Air Track

    Inclined Plane – Simple Machines

    Inclined Plane – Sliding Objects

    Free Fall Tower

    Potential Energy on Shelves

    Uniform Circular Motion

    Orbital Motion – Kepler's Laws

    Moment of Inertia

    Gravity Pitch

    Gravitational Force

    Free Fall Laboratory

    Weight and Mass

    Determining a Spring Constant

PhET.colorado.edu

    Force and Motion: Basics

    Friction

    Gravity and Orbits

    Gravity Force Lab

    Gravity Force Lab – Basics

    Vector Addition

- Balancing Act
- Forces and Motion
- Forces and Motion Basics
- Forces in One Dimension
- Friction
- Lunar Lander
- Masses and Springs
- Masses and Springs Basics
- Ramp: Forces and Motion
- The Ramp
- Torque
- Molecular Polarity
- Atomic Interactions

Physicsclassroom.com

Physics Interactives

Newton's Laws of Motion

- Balanced vs. Unbalanced Forces
- Force
- Free Body Diagrams
- Rocket Sledder
- Which One Doesn't Belong?
- Skydiving
- Elevator Ride

Force Free Body Diagrams

Kinematics

    Name That Motion

    Match That Graph

    Graph That Motion

    Graphs and Ramps

    Two-Stage Rocket

    Distance vs. Displacement

    Vector Walk

Concept Builders

  Newton's Laws of Motion

    Balanced vs. Unbalanced Forces

    Force and Motion

    Change of State

    Recognizing Forces

    Match That Free-Body Diagram

    Which One Doesn't Belong? Force and Motion

    Net Force (and Acceleration) Ranking Tasks

    Air Resistance and Skydiving

    Normal Force Card Sort

  Kinematics

    Distance vs. Displacement

    Matching Pairs: Speed – Distance – Time

    Acceleration

    Name That Motion

Motion Diagrams

Graph That Motion

Match That Graph

Position-Time Graphs – Conceptual Analysis

Position-Time Graphs – Numerical Analysis

Velocity – Time Graphs

Dots and Graphs

Words and Graphs

Which One Doesn't Belong?

Free Fall

Up and Down

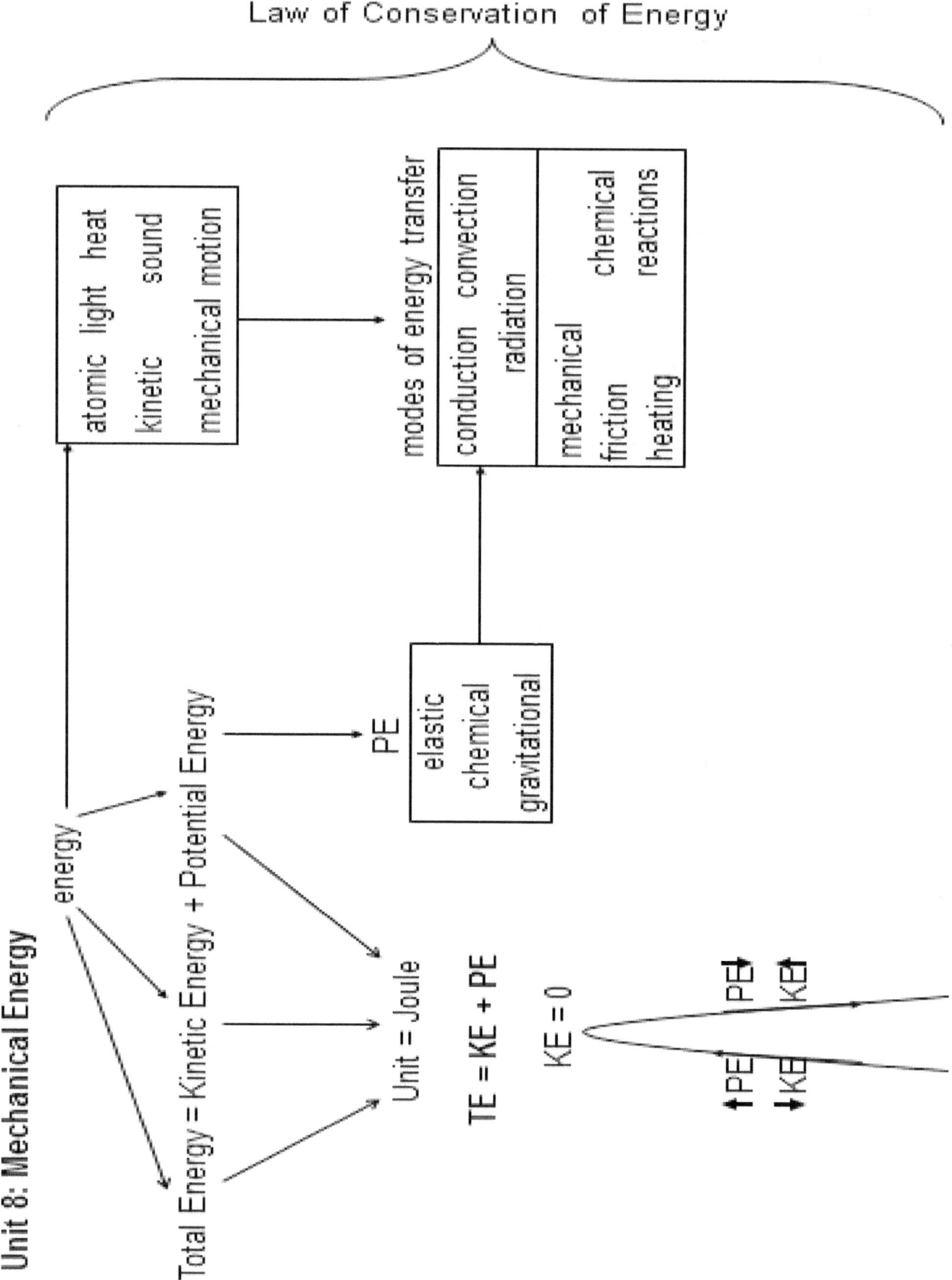

Simple Middle School Physical Science Investigations   Seven Sides Publishing

The Energy of a Pendulum Lab

Directions:

You will need a **string**, hanging **masses (10 g, 20 g, and 50 g)**, a **ring stand**, and a **stopwatch**. **Looking at the materials and lab we will be using, what are the safety precautions we should take to protect ourselves and materials during the investigation?**

1) Take one end of the string and tie it to one side of the ring on a ring stand set as high as you can. Have enough string to where you tie it to the other side of the ring; when you hook the largest hanging mass in the middle of the string, it will be able to swing freely just above the ground/table.
2) Measure the pendulum's length by measuring the distance between the ring's front to the bottom of the mass. Write this in the Data Table 1.
3) Have the pendulum swing back and forth and time how long it takes to go through 15 swing cycles with a 50 g mass on it. Write this in Data Table 1.
4) Repeat the procedure in #3 with a 20 g mass on it.
5) Repeat the procedure in #3 with a 10 g mass on it. Write this data in both Data Table 1 and the first part of Data Table 2.
6) Now wind the string around the front of the ring so you shorten the pendulum's length. Measure its length as you did in step #2. Write this measurement in Data Table 2.
7) Repeat the procedure in #3, keeping the 10 g mass on, and write that data in Data Table 2.
8) Repeat this procedure in #6-7 one more time for a shorter distance. Write your data in Data Table 2

Data Table 1

Mass (g)	Length of Pendulum (cm)	Time for 15 Cycles
50 g	cm	s
20 g	cm	s
10 g	cm	s

Data Table 2

Mass (g)	Length of Pendulum (cm)	Time for 15 Cycles
10 g	cm	s
10 g	cm	s
10 g	cm	s

Questions:

1) Where in the swing is the pendulum moving the fastest?

2) Where in the swing is the pendulum moving the slowest?

3) Potential Energy is defined as having the ability to do work. Where do we see the pendulum with the highest Gravitational Potential Energy in the swing?

4) Kinetic Energy is the energy of movement. Where does the pendulum have the highest kinetic energy in the swing?

5) How does changing the mass affect the time it took to complete 15 cycles of the pendulum?

6) How does changing the pendulum length affect its time to complete 15 cycles?

7) How do the PE and KE change as the pendulum moves from its highest point to its lowest point?

8) How do the PE and KE change as the pendulum rises from its lowest point to its highest point?

9) Write a formula showing the relationship between Potential Energy (PE), Kinetic Energy (KE), and Total Energy (TE).

10) How does this information relate to the energy change when kicking a ball straight up into the air?

Simple Middle School Physical Science Investigations   Seven Sides Publishing

Energy and Rockets Lab

Materials:

You will need a **stomp air rocket launcher**, **2 lb.** and **8 lb. medicine balls,** and measure your **rocket's** mass with a **scale. Looking at the materials and lab we will be using, what are the safety precautions we should take to protect ourselves and materials during the investigation?**

Rocket's mass: _____

Prediction:

1) If you drop the same medicine ball from a high height and a low height, which height should make the rocket fly higher?

Experiment:

2) Drop the same medicine ball from two significantly different heights on the stomping part of the rocket launcher. Which height made the rocket fly the highest?

Conclusion:

3) Why did that height make the rocket fly higher than the other?

Prediction:

4) Which will cause the rocket to fly higher if dropped at the same height, the 2 lb. medicine ball or the 8 lb. medicine ball?

Experiment:

5) Using the same rocket launcher, drop each medicine ball on the rocket launcher from the same height and see which ball caused the rocket to fly higher. Which ball caused the rocket to fly higher?

Conclusion:

6) Why did that medicine ball cause the rocket to fly higher?

7) What type of energy was in the medicine ball before you dropped it?

8) What type of energy was in the medicine ball when it hit the rocket launcher?

9) What type of energy did the rocket get when it took off?

10) What type of energy did the rocket have at its highest point?

11) Describe the energy transformations from the dropped medicine ball and how it transformed into the rocket.

Analyzing Elastic Potential Energy

Directions:

You will need a **rubber band**, a **meter stick**, a **coin**, or some type of **disc. Looking at the materials and lab we will be using, what are the safety precautions we should take to protect ourselves and materials during the investigation?**

1) Make marks on your table every .5 cm for 4 centimeters.
2) Place the rubber band on the zero mark so that the rubber band has no slack between your fingers.
3) Place your coin or disc in front of that rubber band, pull it back .5 cm, and release it.
4) Measure how far the disc traveled. Put this data in Data Table 1 below.
5) Repeat the procedures in #s 2-4, pulling the disc back 1 cm this time, and place this data in Data Table 1 below. Repeat this at .5 cm intervals longer until you pull it back 4 cm.

Data Table 1

Length Pulled Back (cm)	Distance Traveled (cm)
.5 cm	
1 cm	
1.5 cm	
2 cm	
2.5 cm	
3 cm	
3.5 cm	
4 cm	

Questions:

1) How did the disc's takeoff speed seem to change as you pulled the disc back farther and farther?

    a. What does this imply about the kinetic energy of the disc when released?

    b. What does this imply about the potential energy as you pull the rubber band back?

2) How does this investigation show the conservation of energy?

3) What causes the disc to stop?

4) Where does that energy go as it stops?

5) How does this relate to slowing down spacecraft that renter the atmosphere?

    a. What precautions do we have to take?

Happy and Sad Balls

Directions:

You will need a **scale** and a **pair of rubber balls**; one that is **happy (bounces)** and the other that looks just like it that is **sad (does not bounce)**. You will also need a **meter stick** and a **plastic shoebox** with a **lid. Looking at the materials and lab we will be using, what are the safety precautions we should take to protect ourselves and materials during the investigation?**

1) Find the mass of both balls: Happy ball _____kg Sad ball _____kg
2) Hold the meter stick vertically on the table so that 0 cm is on the table and 100 cm is in the air.
3) Hold the happy ball one meter off the table next to the meter stick and drop it so that it bounces next to the meter stick. Measure the height of how high it bounces. Write this in Data Table 1.
4) Repeat the procedure in #3 two more times and enter the data in Data Table 1.
5) Now place the box on the table and the meter stick on the box. Hold the happy ball one meter above the box next to the meter stick, drop it so that it bounces off the box, and measure how high the ball bounces. Write this information in Data Table 1.
6) Repeat the procedure in # 5 two more times and write that data in Data Table 1.
7) Now take the sad ball and repeat the procedures in #s 2-3, bouncing it on the table and write that data in Data Table 1.
8) Repeat #7 two more times and write the data in Data Table 1.
9) Calculate the bounce trials' average height in Data Table 1 by adding the three bounces and dividing by 3 for each ball to surface set-up.

Data Table 1

Ball	Surface	Trial 1 Bounce (m)	Trial 2 Bounce (m)	Trial 3 Bounce (m)	Average Height (m)
Happy	Table				
Happy	Box				
Sad	Table				

Questions:

1) What happens to its gravitational potential and kinetic energy as the ball falls?

2) What happens to their kinetic energy when they hit the floor?

3) How did elastic potential energy help it bounce?

4) How did the bounce heights compare when dropped on the box instead of the floor?

5) Why did this happen? Did you hear anything or see anything move?

6) Why do you think the sad ball did not bounce as high?

7) Did the sad ball have much energy transfer into elastic potential energy?

8) How does the information from this lab help us understand how bulletproof vests work?

Conservation of Energy in a Toy

Directions and Questions:

What you are going to do today is called reverse engineering. You will look at **toys using simple machines, see how they work, and trace how energy is transferred through the toy from when it enters it until it leaves it**. You will describe what simple machines or devices are used to make the toy move for its function. **Looking at the materials and lab we will be using, what are the safety precautions we should take to protect ourselves and materials during the investigation?**

1) Describe in as much detail as you can what your toy does.

2) How does the energy enter your toy?

3) Describe how the energy is transferred through your toy (tell what machines or devices are used).

4) Tell all the different ways energy leaves your toy.

5) Leave your toy with this paper and look at someone else's toy and see how accurate they were, and add observations and corrections as they do the same to yours.

6) Discuss the changes/additions you made to each other's work (this is collaboration).

Simple Middle School Physical Science Investigations — Seven Sides Publishing

The Energy of Colliding Objects

Directions:

You will need a **scale**, **masking tape**, some type of **ramp** or **track**, a **meter stick**, a **ping pong ball**, a **tennis ball**, a **baseball**, and a **softball**. Looking at the materials and lab we will be using, what are the safety precautions we should take to protect ourselves and materials during the investigation?

1) Find the mass of each object and write them in Data Table 1.

2) Place a piece of tape on the floor and line your ramp/track up to it.

3) Place the softball at the bottom of the track on the tape.

4) Roll the ping pong ball down the track and measure how far it moved the softball.

5) Roll the tennis ball down the ramp and measure how far it moves the softball.

6) Roll the baseball down the ramp and measure how far it moves the softball.

Data Table 1

Ball	Mass (g)	How far it moved the softball (cm)	Rank Kinetic Energy
Ping Pong			
Tennis			
Baseball			

Questions:

1) Which ball moved the softball the farthest?

2) This ball had the most kinetic energy and passed on to the softball. Rank the KE of the balls at the bottom of the ramp from highest to lowest in Data Table 1.

3) Which ball had the most Gravitational Potential Energy at the top of the Ramp?

4) In this experiment, what caused the biggest transfer of energy?

5) How else could we transfer more energy?

6) Test it to see if it works.

Who's got the Power?

Directions:

You will need a **dowel** or **PVC pipe**, a **500 g mass**, **string** about 1m long, **masking tape**, and a **stopwatch. Looking at the materials and lab we will be using, what are the safety precautions we should take to protect ourselves and materials during the investigation?**

1) Tie and tape one end of the string to the dowel/PVC pipe. Measure 75 cm down and put some masking tape as a marker. Hook the 500 g mass to the other end of the string.

2) Take turns to roll the dowel/PVC pipe to 75 cm, pulling the mass up while timing how long it takes with the stopwatch.

Data Table

Measurement	Student 1	Student 2
Time (s)		
Force (N)		
Distance (m)		

Questions:

1) Calculate the work both students did. (W = F x d)

2) Calculate the power of both students. (P = W/t)

3) Compare the amount of work each of you did. Why is the work the same?

4) Compare the power of both students. Who had the most power?

5) Why would the powers differ?

6) How does speed affect the amounts of work and power (trick question)?

7) How does work relate to energy?

8) How does power relate to energy?

Simple Middle School Physical Science Investigations   Seven Sides Publishing

Levers Lab

Directions Part 1:

You will need a **meter stick**, **string**, **hanging masses**, a **fulcrum collar**, and a **support stand** for the fulcrum. Make sure the middle of the fulcrum is lined up on the 50 cm mark. **Looking at the materials and lab we will be using, what are the safety precautions we should take to protect ourselves and materials during the investigation?**

1) Take a 50 g mass and some string and hang it 20 cm from the fulcrum. Then hang a 25 g mass on the other end of the meter stick at a distance that allows it to balance the meter stick on the stand with the fulcrum. Use that distance to fill in Data Table 1 for Trial 1.
2) Now move the 50 g mass 30 cm away from the fulcrum. Now take a 100 g mass and place it on the other side of the meter stick at a distance that balances the meter stick again. Fill in Data Table 1 with the measurement for Trial 2.
3) Now take two different masses and line them up on opposite sides of the meter stick to get those to balance the meter stick two more times. Write this information in Data Table 1.

Data Table 1

Trial	Mass on Left	Distance on Left	Mass on Right	Distance on Right
1	50 g	20 cm	25 g	
2	50 g	30 cm	100 g	
3				
4				

Questions Part 1:

1) What pattern do you see with the mass and distance from the fulcrum?

2) Write a formula that tells you how to balance a meter stick with two different masses.

3) If you convert the masses to weight or force, you can find the torque. How would you convert these masses to force?

4) Calculate the force of weight for each of the masses.

5) Now use the formula Torque = force x distance to calculate the torque on each side of the meter stick for each trail.

6) What do you notice about the unit for torque? Where is it also used?

7) How do you move a force to achieve a mechanical advantage?

Directions Part 2:

1) Now take three different masses and balance two of them on the left and one on the right. Write the masses and distances in Data Table 2.
2) Repeat Part 2 procedure #1 three more times with different masses and distances. Write your data to fill in Data Table 2.
3) How do you think the left side will compare to the right?

4) Write a formula for this equality.

Data Table 2

Trial	1st Mass Left	1st Distance Left	2nd Mass Left	2nd Distance Left	Mass on Right	Distance on Right
1						
2						
3						
4						

Questions Part 2:

1) So does your formula work with the results?

2) If not, write a formula that will work with Data Table 2.

3) Use this formula to calculate and prove the equality mathematically for trial 1.

4) Prove the equality for trial 2.

5) Prove the equality for trial 3.

6) Prove the equality for trial 4.

Simple Machines Lab

Directions and Questions:

You will need a **shoe**, a **cart with wheels**, a **screwdriver**, a **round doorknob**, a **lever doorknob**, and a **dual-range force sensor** attached to an **interface** connected to a **computer** with **Logger Pro. Looking at the materials and lab we will be using, what are the safety precautions we should take to protect ourselves and materials during the investigation?**

1) A wheel and axle is a modified lever that allows you to move a force over a longer distance. The bigger the radius, the bigger the mechanical advantage. (**Work = F x d**)
2) Hook your force sensor up to a shoe and slide it across the table to see the average force applied by the force sensor you are pulling on the shoe. What was the force?

3) Now do the same thing as #2 but put the shoe on the cart, so it rolls across the table. Which trial required the least amount of force?

   a. Which force is greater, sliding friction or rolling friction?

4) Which trial did more work?

5) Which trial allowed you to do work easier? That is because of mechanical advantage, allowing you to do the same amount of work but easier over a bigger distance.

6) Look at the **W = F x d** equation. If the work stays the same, what happens to the force if you move it over a bigger distance?

7) Compare a lever doorknob and a circular doorknob. How are the doorknobs alike?

8) How are the doorknobs different?

9) How do you get a mechanical advantage to open a door with the lever?

10) How do you get a mechanical advantage with the round doorknob to open a door?

11) How can a screwdriver give you a mechanical advantage?

12) Which screwdriver would make it easier to screw in a screw, a thick handle or thin?

13) How would a ramp give you a mechanical advantage? Draw it out and explain it below.

14) Explain how the forces on a wedge under a door hold the door back. Draw a diagram and explain it below.

15) How did all of the simple machines make work easier?

16) How does a clothespin/clip work as a simple machine?

17) Find three other examples of simple machines and tell how they work to achieve a mechanical advantage.

Pulley Lab

Directions and Questions:

You will need a **set of hooked masses**, at **least two pulleys**, some **string**, a **ring** clamped onto a **ring stand**, and a **dual-range force sensor** attached to an **interface** connected to a **computer** with **Logger Pro**. Looking at the materials and lab we will be using, what are the safety precautions we should take to protect ourselves and materials during the investigation?

1) A pulley is a wheel and axle, a modified lever that can change the force's direction or increase the mechanical advantage to do work with less force. They can be a single fixed pulley, movable pulley, multiple fixed pulleys, or a block and tackle with a movable pulley.

2) Set up a single fixed pulley on the ring stand. Run a string through the pulley and connect it to the mass. Attach the dual-range force sensor to the other end of the string and pull down, measuring the force required to lift the mass. What was the mass that you picked up?

3) What is the force of the mass (m x 9.81)?

4) What was the force the sensor showed you needed to lift it?

5) Why do you think there was there a small difference in the forces?

6) As you pull down, what direction does the weight go?

7) Now set up a single movable pulley by tying one end of the string to the ring stand and running the string through the pulley that is hooked to the same mass, and the other end of the string is fixed to the force sensor. What is the force lifting the mass this time?

8) How does this compare to #4?

9) Now set up a block and tackle with one fixed pulley and one movable pulley like the one shown in **Figure 1** below. Make sure the force sensor is pulling like the arrow on the left.

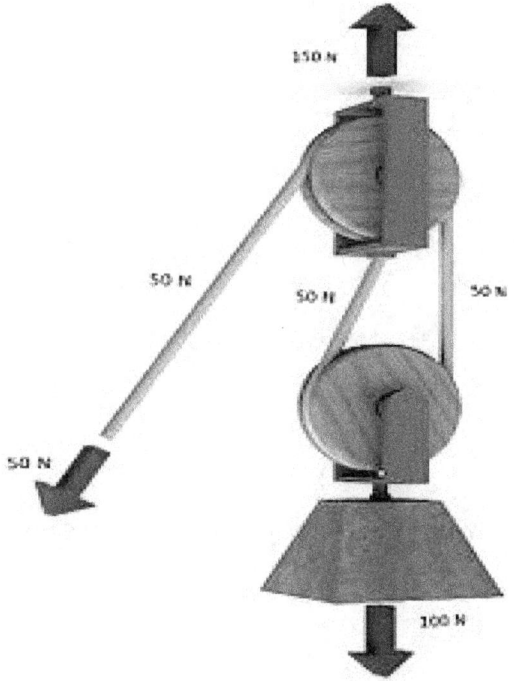

Figure 1: Block and Tackle picture from Creative Commons via Wikipedia.

10) What is the force showing from the sensor?

11) How does it compare to #s 4 & 8?

12) How do you think we achieved a mechanical advantage in both 8 and 10?

13) If you have multiple fixed pulleys, try to set up a system with a higher mechanical advantage. When you do, draw a diagram below showing how you achieved that higher mechanical advantage. Make sure you include what the mechanical advantage was and why.

14) What are some pulleys that we use in our lives for work and recreation?

Simple Middle School Physical Science Investigations   Seven Sides Publishing

Bicycle Lab

Directions:

You will need a **gear-changing bicycle**. Flip the bike upside down, resting on the seat and handlebars. Have the bicycle in the lowest gear. **Looking at the materials and lab we will be using, what are the safety precautions we should take to protect ourselves and materials during the investigation?**

1) Pedal the bike and notice the speed of the back wheel. Now shift it to a higher gear, pedaling the bike at the same pace. How fast is the back wheel moving now compared to when it was in a lower gear?

2) This increase in speed was due to a speed advantage. How do you think that happened?

3) Gears allow us to choose to move vehicles slowly with a high mechanical advantage at a low-speed advantage or fast with a low mechanical advantage but with a high-speed advantage. How do you think this is done while driving the vehicle?

4) Look at the gears around the cranks of the bicycle. Count how many teeth there are around each ring (starting with the biggest gear and ending with the smallest). Write this in Data Table 1.
5) Look at the gears around the back axle of the bicycle. How many teeth are on each gear (starting with the smallest and ending with the biggest)? Write them in Data Table 1.
6) Calculate the speed advantage by dividing the front gear by the back gear. Fill in the data table for the combination gears shown in each row.
7) Calculate the mechanical advantage by dividing the back gear by the front gear. Fill in the data table for the combination of gears shown in each row.

Data Table 1

Speed	Front Gear # of Teeth	Back Gear # of Teeth	Speed Advantage	Mechanical Advantage
Fastest	Biggest:	Smallest:		
	Biggest:	2nd Sm:		
	Middle:	3rd Sm:		
	Middle:	3rd Big:		
	Smallest:	2nd Big:		
Slowest	Smallest:	Biggest:		

Questions:

1) How is the mechanical advantage for the fastest speed compared to the mechanical advantage for the slowest speed?

2) How is the speed advantage for the fastest speed compared to the speed advantage for the slowest speed?

3) We only looked at the combination of gears for six speeds. How many speeds does your bike have?

4) Why are these bikes built with so many gears/speeds?

5) When would you use the lowest speed/gears?

6) When would you use the highest speed/gears?

7) Gears are a modified wheel and axle, which is a modified lever. You could even say that the gears set up on the bike are a modified pulley system. What are some other simple machines that are on this bicycle? Tell what they are and how they function on the bike.

8) How does the speed of the bicycle relate to energy?

9) What kind of force do we need to have high kinetic energy?

Virtual Investigations that go with Mechanical Energy

ExploreLearning.com:

    Sled Wars Gizmo

    Roller Coaster Physics Gizmo

    Inclined Plane – Sliding Objects Gizmo

    Energy of a Pendulum Gizmo

    Air Track Gizmo

    Trebuchet Gizmo

    Potential Energy on Shelves Gizmo

    Inclined Plane – Simple Machine Gizmo

PhET.colorado.edu:

    Energy Forms & Changes

    Energy Skate Park

    Hook's Law

    Masses and Springs

    Pendulum Lab

    The Ramp

Physicsclassroom.com:

    Physics Interactives:

        Work and Energy

        It's All Uphill

        Stopping Distance

        Roller Coaster Model

        Chart that Motion

Vibrating Mass on a Spring

Concept Building:

Work and Energy

Name That Energy

What's Up (and Down) with KE and PE?

Energy Ranking Tasks

Work

Match That Bar Chart

LOL Charts (a.k.a., Energy Bar Charts)

Unit 9: Thermal Energy

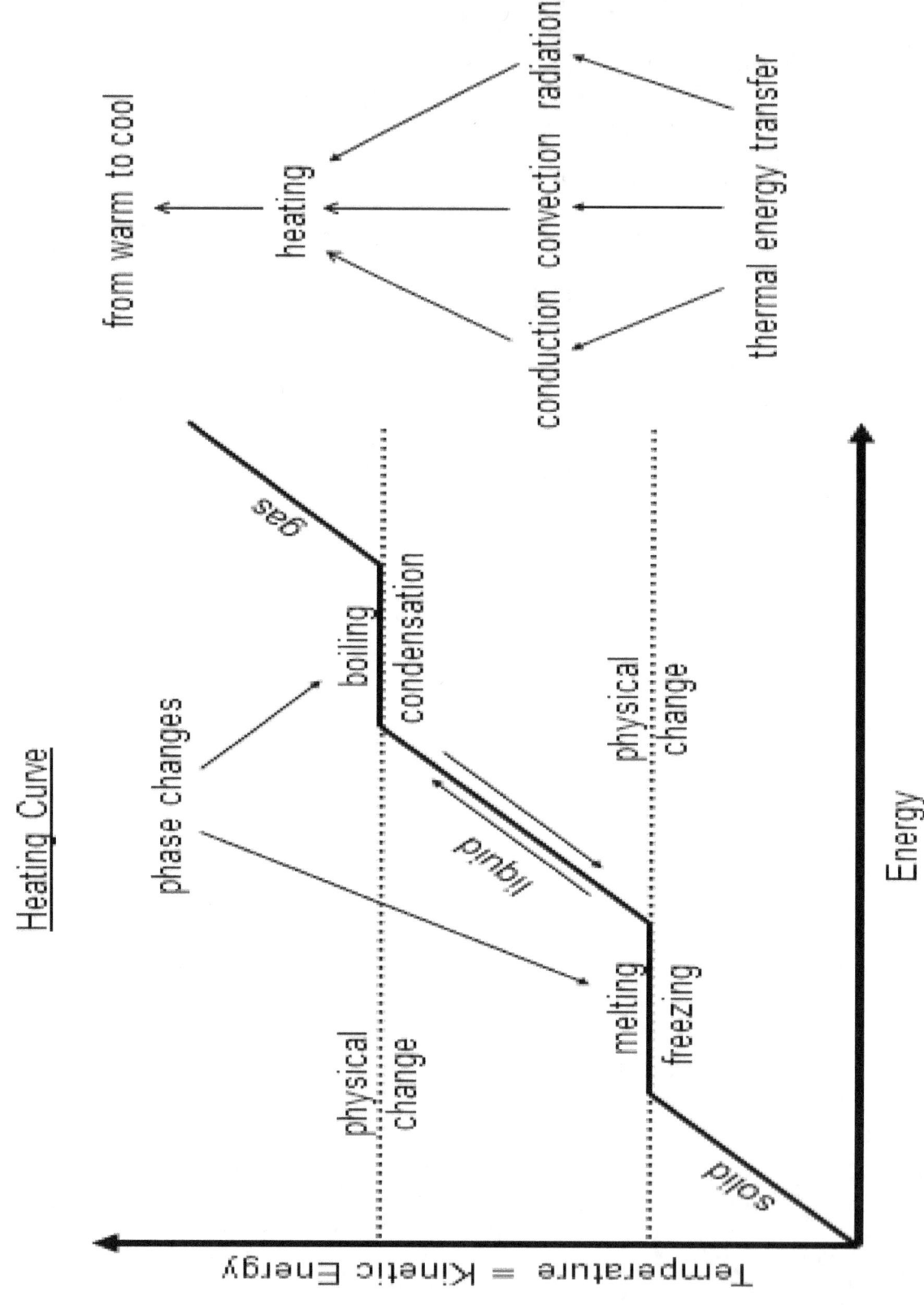

Energy Transformation Balls

Directions:

You will need a pair of **steel energy transformation balls**, an **index card** or piece of **paper**, and **safety goggles. Looking at the materials and lab we will be using, what are the safety precautions we should take to protect ourselves and materials during the investigation?**

1) Put on your safety goggles. Have one person take the steel energy transformation balls and hold one in each hand. Have another person vertically hold out a piece of paper or an index card. The person holding the energy transformation balls should then smash the two balls together on the paper with a very strong force. **Make sure not to get anyone's fingers in the way.**

2) Observe what happens to the paper where the balls hit and what you smell.

3) Discuss with your teacher the questions that follow.

Questions:

1) What do you see on the paper where the balls hit?

2) What do you smell?

3) What do you think happened?

4) How was energy transformed?

5) How is this collision like a meteor hitting the Earth?

6) How does this relate to the Theory of Relativity $E = mc^2$?

7) How does this relate to the Big Bang Theory?

Simple Middle School Physical Science Investigations — Seven Sides Publishing

Convection in Liquids and Gases

Directions:

Fill a **large beaker** with **water**, add **pepper** to it, place it on a **hotplate,** and heat the water to just below the boiling point with your **safety goggles** on. **Looking at the materials and lab we will be using, what are the safety precautions we should take to protect ourselves and materials during the investigation?**

1) Draw a picture of the motion of the pepper in hot water:

2) Describe the motion you see in hot water:

3) What do you think is causing this motion (go into detail)?

4) Light a **candle** with a **match/lighter** and gently blow it out. Which direction does the smoke go?

Questions:

1) Describe how the particle of the pepper moved as the water became hotter.

2) Describe how the particles of pepper moved as the water became colder after losing heat on the surface.

3) Explain how convection currents formed in the beaker.

4) Explain why the motion of the particles changed as the burner heated up the water.

5) Explain how the thermal energy moved through the beaker of water.

6) Which direction did the smoke go when the candle was blown out?

7) Explain why the smoke went in that direction.

Observing Conduction Convection and Radiation

Directions and Questions:

You will need **safety goggles**, a **hotplate**, **Jiffy Pop popcorn**, a **hot air popper, unpopped popcorn, microwave popcorn**, and a **microwave oven. Looking at the materials and lab we will be using, what are the safety precautions we should take to protect ourselves and materials during the investigation?**

1) Heat popcorn in the Jiffy Pop skillet on a hotplate. Once the popcorn gets hot, what happens to it?

2) Describe how the energy moved from the hotplate to the popcorn in as much detail as you can.

3) How do you know the energy got there?

4) What method of heat transfer was this?

5) Put popcorn in a running hot air popcorn popper. What happens when the popcorn gets hot?

6) Describe how the energy moved from the popper to the popcorn in as much detail as you can.

7) How did you know the energy got there?

8) What method of heat transfer was this?

9) Microwave a bag of popcorn. What happens to the popcorn?

10) Describe how the energy got to the popcorn in as much detail as you can.

11) How do you know the energy got there?

12) What method of heat transfer was this?

13) What direction did the thermal energy move as the popcorn was heated and popped?

Simple Middle School Physical Science Investigations   Seven Sides Publishing

The Direction Thermal Energy Moves

Directions and Questions:

Place ice cubes in a beaker of water. Looking at the materials and lab we will be using, what are the safety precautions we should take to protect ourselves and materials during the investigation?

1) Thermal energy flows from high energy to low energy. In other words, it moves from warmer areas to colder areas. Write down your observation of what happens to the ice cubes.

2) What was the source of thermal energy?

3) How did the thermal energy move to cause the ice to melt?

4) What gained thermal energy in this system?

5) What lost energy in this system?

6) When it is cold outside, and you open the door to a heated house, in which direction will the energy flow?

   a. Is this accurate when your parents tell you to close the door because you are letting the cold air in? Explain why.

7) When it is warm outside, and you open the door to a cooled house, which direction is the thermal energy flowing?

   a. Is it accurate when your parents tell you to close the door, saying you are letting the cold air out? Explain why.

Observing Molecular Motion

Directions:

You will need at least three of the **biggest beakers** in your school. They all will be filled with water. One beaker **filled with water** needs to be put in a **refrigerator** so the water is cold. You will also need one beaker on a **hotplate** before doing the demo, so it heats up (if it boils, you can turn the heat off). The third beaker will be room temperature water straight out of the tap. Lastly, you will need some **food coloring. Looking at the materials and lab we will be using, what are the safety precautions we should take to protect ourselves and materials during the investigation?**

1) Line all three beakers up from coldest to warmest where the whole class can see. Place a drop of food coloring in the cold beaker and have the students watch how the food coloring spreads.
2) Put a drop of food coloring in the room temperature water, then in hot water on the hotplate. Have students observe the movement in all three beakers. It will not take long for the hot one to become homogeneous.
3) Have the students draw what they see in the three beakers below.

    Cold   Warm   Hot

Questions:

1) In which beaker did the dye move the fastest?

2) In which beaker did the dye move the slowest?

3) Why do you think the dye moved at different speeds?

4) Temperature is defined as the average kinetic energy of molecules. How does this explain what was happening in the hot, warm, and cold water?

5) Which beaker had the most energy?

6) How can this explain why we get hurt when we touch something hot?

Seeing the Heating Curve (Thermal Energy)

Directions:

You will need a **beaker**, **frozen water**, and a **temperature probe** suspended in it; this needs to be prepared for the day before in the **freezer**. Freezing water in the beaker many times breaks the beakers, so freeze the water in paper or **Styrofoam cups** while suspending the temperature probes in the water. When doing the investigation, peel off the cup and put the ice in the beaker. Attach the probe to an **interface** connected to a **computer** with **Logger Pro**. You will need to place the beaker of ice on a **hotplate**. You also need a **ring stand** and **clamp** to hold the temperature probe up as the ice melts. **Looking at the materials and lab we will be using, what are the safety precautions we should take to protect ourselves and materials during the investigation?**

1) Make sure the Logger Pro is set up to collect temperature data every second for at least 20 minutes.
2) Click "Collect" and turn the hotplate on high.
3) Watch the setup and data until the ice melts to water and the water boils for a little while. Then click "Stop."
4) Plot the graph Logger Pro gave you in Graph 1 and answer the questions below.

Questions:

1) What was the temperature range of the ice in the data? Change in temperature shows kinetic energy is changing for water.

2) What did the graph look like while the water was ice (before it started to melt)? As the temperature changed, so did the kinetic energy.

3) What does the graph look like as the ice melted (changed from solid to liquid)? This pattern shows potential energy is changing.

4) Is the kinetic energy changing during this phase change?

5) How did the thermal energy move in this system?

6) What was the temperature range of the water after the ice melted and before it boiled? This temperature change is the change in kinetic energy.

7) As the water boiled, what was the temperature?

8) What does the graph look like during this phase change? Potential energy is changing again, but the kinetic energy is not changing.

9) What do you think the temperature range would be for steam?

Graph 1

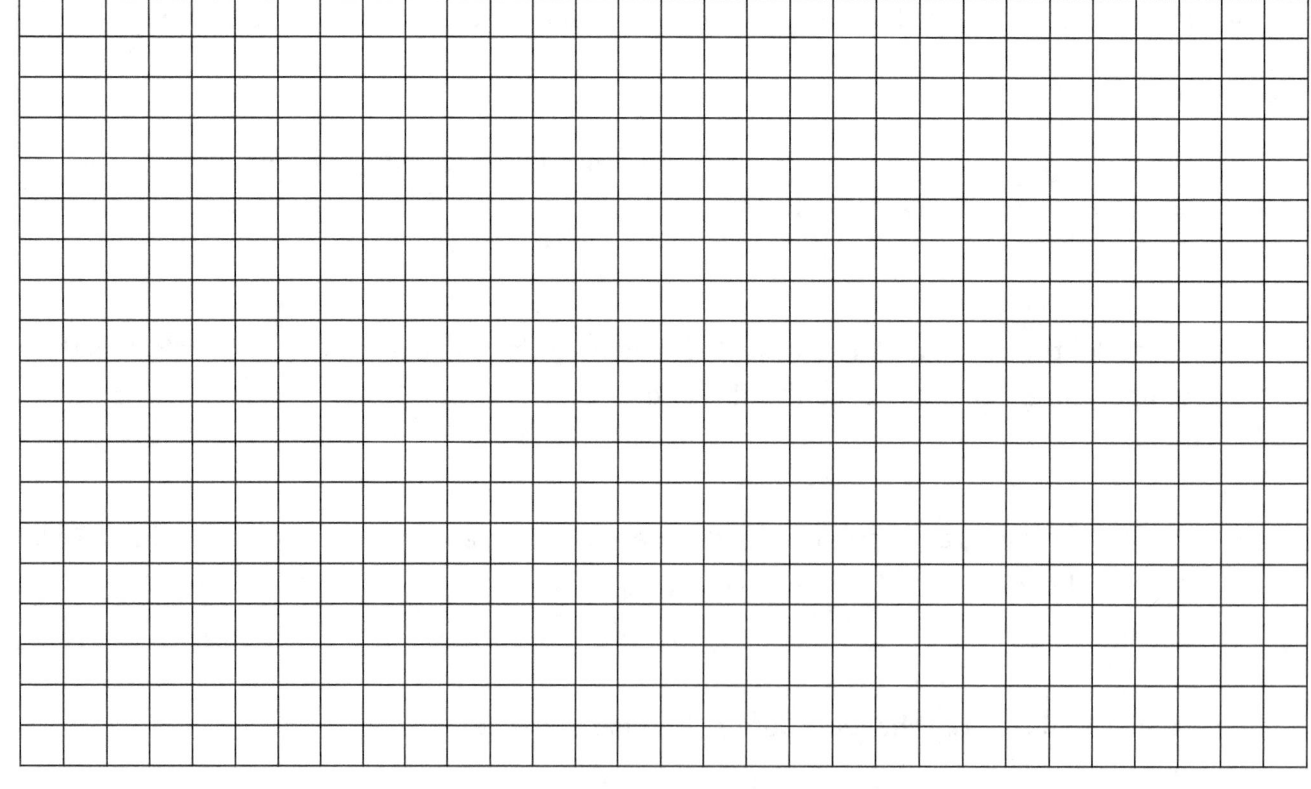

Time (min)

Solar Oven

Directions:

Build a solar oven using a **pizza box, aluminum foil, plastic wrap,** and a **black plate** to collect light and build up thermal energy to cook **food of your choice. Looking at the materials and lab we will be using, what are the safety precautions we should take to protect ourselves and materials during the investigation?**

1) Place the food you will cook on the black plate. Place one piece of aluminum foil shiny side up, covering the inside of the pizza box.
2) With the lid up, cover the top of the box with plastic wrap (leave it loose until you get outside). You may need to tape the back of the box by the lid to the sides so air cannot circulate out.
3) On the pizza box lid, line the inside of the lid with more aluminum foil, shiny side up.
4) Take your food and box outside on a sunny day, put the plate of food in the pizza box, seal the plastic wrap, and set your box down where your box can get exposed to the sun.
5) Angle the lid and the box so the light reflects off the foil into the box using your teacher's instructions. Leave it there and come back later to see if the food is cooked.
6) Eat the food if your teacher permits.

Questions:

1) Why did we line the box with foil?

2) Why did we reflect light on the food with the foil lid?

3) Why did we put plastic wrap over the food, sealing the pizza box?

4) Where was the source of the thermal energy?

5) How was heat transferred into the box?

6) How was the thermal energy trapped in the box to cook the food?

7) How was this like our atmosphere traps thermal energy on the Earth?

Testing the Rate of Heat Movement

Directions:

You will use a **drinking glass, styrofoam cup,** and a **tumbler like Rtic or Yeti** to hold **ice water** poured from a **pitcher.** Measure the temperature change they go through with **temperature probes** in each container connected to an **interface** that is connected to a **computer** with **Logger Pro.** Plot a graph with **colored pencils** on Graph 1. **Looking at the materials and lab we will be using, what are the safety precautions we should take to protect ourselves and materials during the investigation?**

1) Fill a drinking glass, a Styrofoam cup, and a tumbler with equal amounts of water from the pitcher of ice water.
2) Set the time to collect the data for 30 minutes. Place a temperature probe in each container and start collecting the temperatures of each.
3) Plot three lines on the graph for the temperature data in Logger Pro in Graph 1.

Graph 1

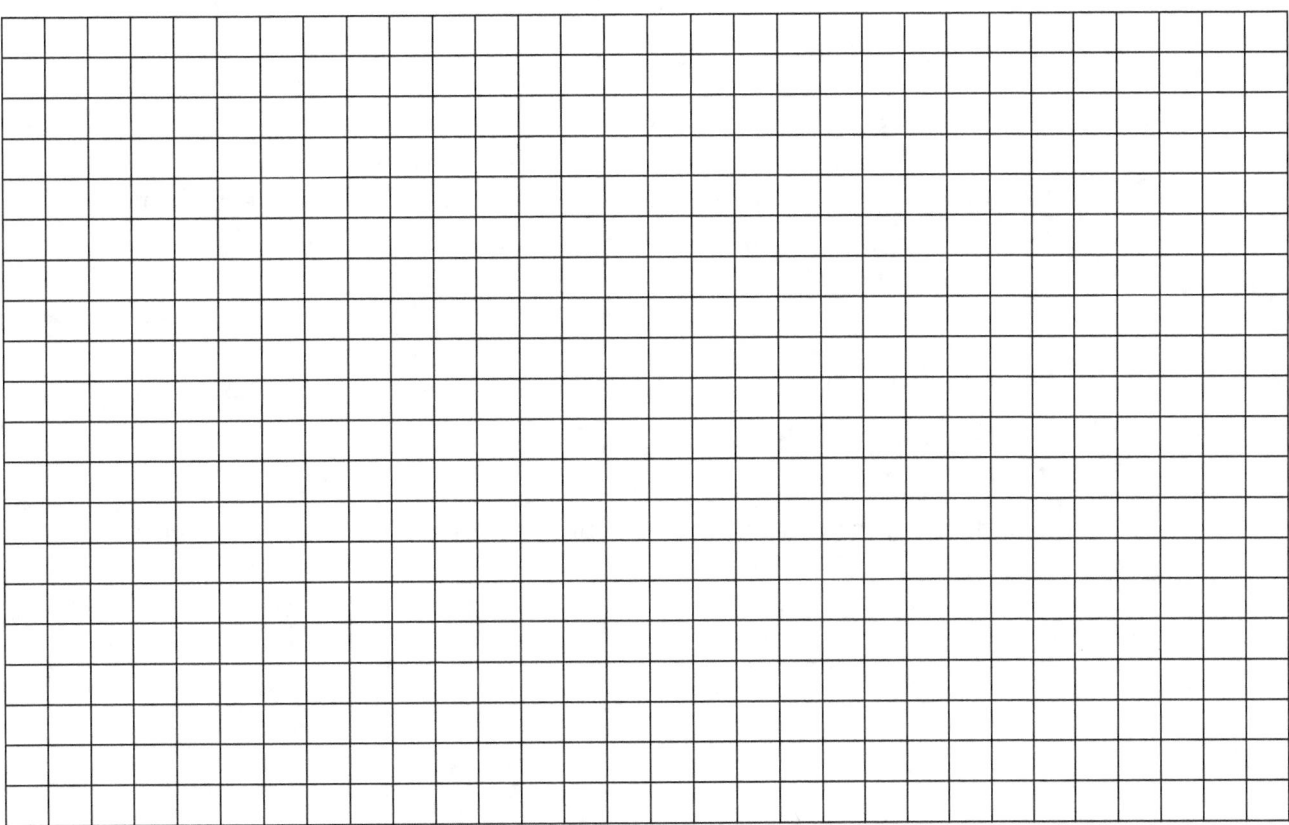

Questions:

1) Why did we use the same amount of water in each container?

   a. Why do you think we did not put ice in the containers?

2) Which container warmed up the fastest?

   a. What do you think about the container that allows heat to transfer in?

3) Which container warmed up the slowest?

   a. What do you think about this container that slowed the movement of heat in, trapping thermal energy out of the container?

4) How were conduction, convection, and radiation involved in this investigation?

Virtual Investigations that go with Thermal Energy

ExploreLearning.com

- Energy Conversions in a System
- Heat Transfer by Conduction
- Heat Absorption
- Conduction and Convection
- Energy Conversions
- Feel the Heat
- Phase Changes
- Temperature and Particle Motion
- Water Cycle
- Phases of Water
- Melting Points
- Chemical and Physical Changes STEM Case
- Chemical and Physical Changes Handbook

PhET.colorado.edu

- Atomic Interactions
- Diffusion
- Energy Form and Changes
- Friction
- Gas Properties
- Gas Intro
- States of Matter
- States of Matter: Basics

physicsclassroom.com/Concept-Builders/Chemistry

Measuring the Quantity of Heat

States of Matter

Unit 10: Electromagnetism

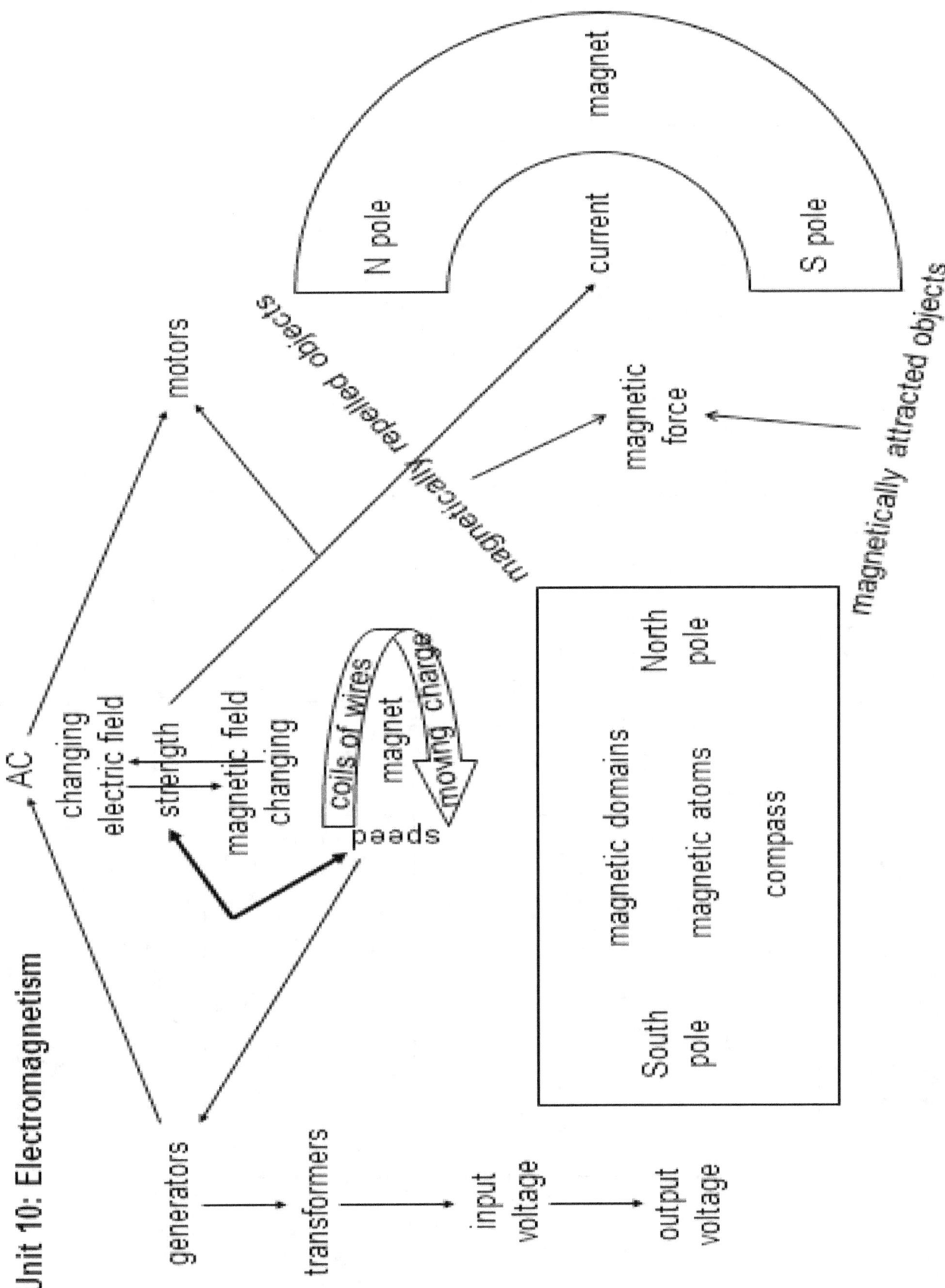

223

Unit 10: Electromagnetism

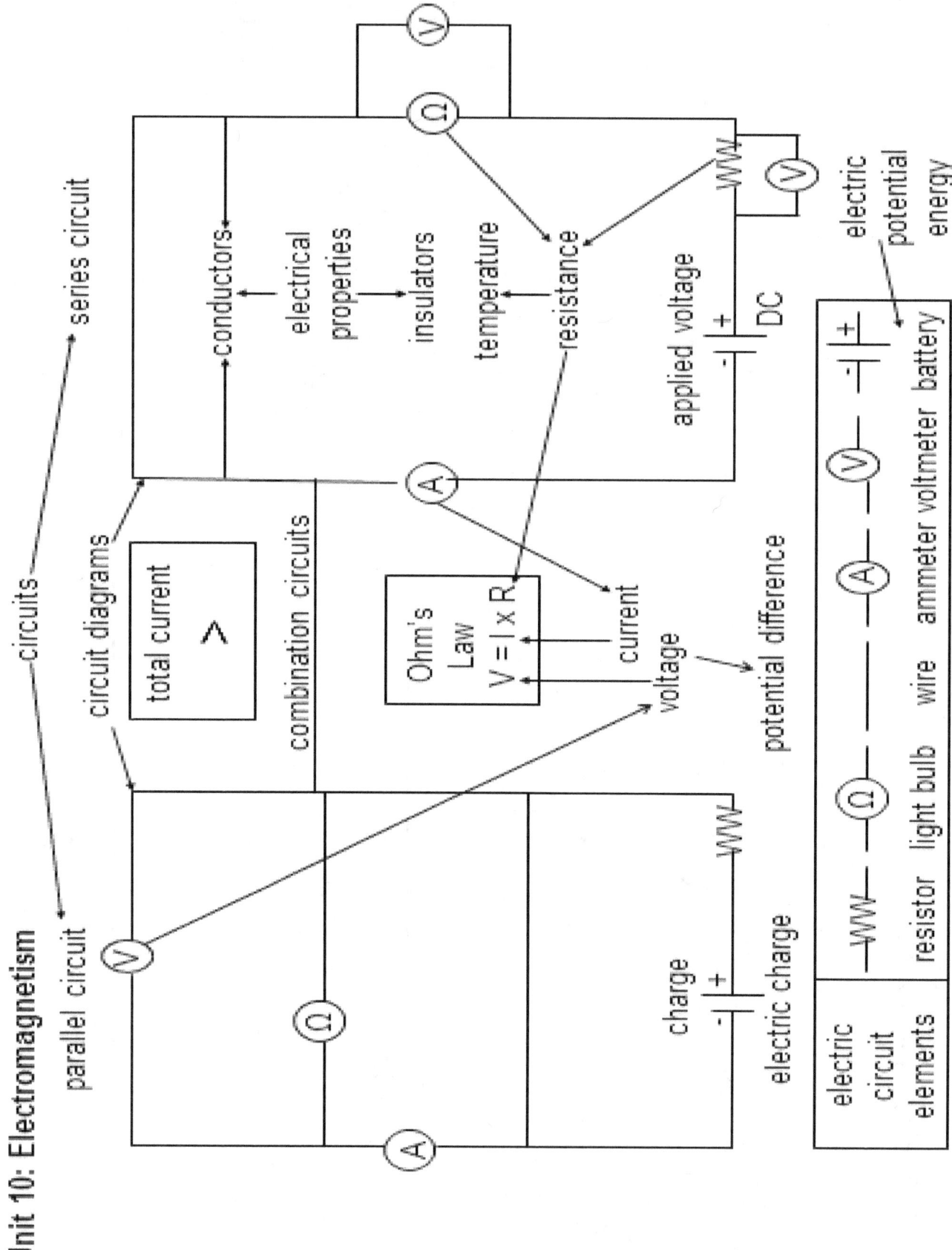

Simple Middle School Physical Science Investigations   Seven Sides Publishing

Static Electricity

Directions and Questions:

You will need a **balloon**, an **electroscope**, a **plastic comb,** some **string,** and a **ring stand**. This lab works best when humidity is low. If you do not have an electroscope, one can be made easily with an **Erlenmeyer flask**, a **rubber stopper** with a hole in it, a **screw or bolt** that fits through the hole, a **paper clip** you wind on the threads of the bolt, and two small pieces of **aluminum foil** you hook on the paper clip that will go inside the Erlenmeyer flask. **Looking at the materials and lab we will be using, what are the safety precautions we should take to protect ourselves and materials during the investigation?**

1) Electrons have a negative charge. Since they are the particles that move outside the atom, these are the ones that we can steal off objects. Take your balloon and rub it in your hair; this steals the electrons from your hair and collects them on the balloon. Now move the balloon away from your hair and back towards it. What happens to your hair as the balloon moves towards it?

2) Why do you think this happens?

3) What charge is your hair?

4) What charge is the balloon?

5) Opposite charges attract, like charges repel. See if the balloon will stick to a wall, your shirt, or under a shelf; if they don't, just rub the balloon in your hair some more. Why do you think the balloon sticks?

6) It is easy to move the electrons from the balloon to metal. Take your charged balloon and move it toward the end of the bolt on the electroscope. What happens to the aluminum foil leaves?

a. Why do you think this happens?

7) Pull the balloon away; what happened?

8) Now touch your charged balloon to the bolt on the electroscope; this should transfer the charge. Move the balloon away. What happened to the leaves of aluminum foil?

9) Now touch your finger to the bolt of the electroscope. What did the leaves do now? Why do you think that happened?

10) Now take the plastic comb and tie it to the ring stand. Run the comb through your hair a few times. Then rub the balloon on your head. Bring the balloon to the comb. How does the comb react to the balloon? Why do you think that happens?

11) Now take the charge away from the comb by touching it with your hand. Then bring the charged balloon back to the comb. How did the comb react this time? Why do you think this happened?

12) If you have a faucet, make a small stream of water come out of it. Bring the charged balloon near it. How does the water react to the charged balloon?

13) Water molecules are polar, like magnets. Why do you think the water reacted that way?

14) Draw a force diagram showing the electrostatic forces repelling the balloon and the comb.

15) Draw a force diagram showing the electrostatic forces attracting the balloon to the comb.

16) What kind of force did we observe today?

The Spinning Match

Directions:

You will need two **nickels**, a **match**, a **clear plastic cup**, and an **inflated balloon. Looking at the materials and lab we will be using, what are the safety precautions we should take to protect ourselves and materials during the investigation?**

1) Take the first nickel and place it on a flat surface, tail up. Take the second nickel and balance it on its side on top of the first nickel.
2) Balance a match on top of the two nickels. Place the plastic cup over the set-up.
3) Rub the balloon on your shirt to collect a negative charge on it.
4) Move the balloon near and around the cup.

Questions:

1) What happens to the match?

2) Why do you think this is happening?

3) What do you think is the charge on the tip of the match?

4) What forces are acting on this set-up?

5) Draw a force diagram of those forces.

Charged Tape

Directions and Questions:

You will need two **scotch tape** pieces about 10 cm in length with one end folded over to make a handle. **Looking at the materials and lab we will be using, what are the safety precautions we should take to protect ourselves and materials during the investigation?**

1) Stick one strip on your desk/table. Take the other strip and stick it on top of the first. Pull both pieces off together, then pull them apart.
2) Bring the non-sticky sides of both pieces of tape together. What do you see happen?

3) Are their charges alike or different?

4) Now place both pieces of tape directly on the table and peel them off.
5) Bring the non-sticky sides together again. What do you see happen this time?

6) Are the charges alike or different on the tape?

7) What do you think caused the change of the tape?

8) When were the electrons stolen?

9) What forces were shown working in this lab?

Seeing Magnets

Directions:

You will need 2 **bar magnets**, a **horseshoe magnet**, and **iron filings** in a **plastic case**. Looking at the materials and lab we will be using, what are the safety precautions we should take to protect ourselves and materials during the investigation?

1) The iron filings align themselves in the magnetic field, allowing us to see the field. Place the bar magnet on top of the plastic case of iron filings and shake them. Draw a picture of the pattern of the iron filings around the magnet.

2) Now place the horseshoe magnet on top of the plastic case of iron filings and shake them. Draw a picture of the pattern of iron filings around the magnet.

3) Now place 2 bar magnets in line so that they are attracted to each other on the plastic case of iron filings and shake them, keeping the two magnets apart. Draw a picture of the pattern of iron filings around the two magnets and arrows showing the forces.

4) Now place 2 bar magnets in line so that they are repelled by each other on the plastic cast of iron filings and shake them, keeping the two magnets apart. Draw a picture of the pattern of iron filings around the two magnets and arrows showing the forces.

Questions:

1) Which seems to have greater strength, the bar magnet or horseshoe?

2) How do you know?

3) How many humps do you see with the attracted magnets?

    a. Is the field connecting or separating between the two magnets?

4) How many humps do you see with the repelling magnets?

    a. Is the field connecting or separating between the two magnets?

5) How does the magnetic field around the attracted magnets appear different than the repelling magnets?

6) Take a bar magnet and tie a string to the middle of it. Hang the magnet away from anything that could attract it. Watch how it orients itself. Place an "N" on the side facing north; this is the magnet's north side. Spin it a little bit and watch it again orient itself. What did you notice?

7) What else is a magnet?

8) Is the North Pole on the Earth magnetically north or south? How do you know?

Making Electromagnets

Directions:

You will need two **dry-cell batteries**, a **battery pack**, a stronger **wet-cell battery**, two identical **bolts**, thinly insulated **wire**, **BBs** in a **container**, and an **empty container**. **Looking at the materials and lab we will be using, what are the safety precautions we should take to protect ourselves and materials during the investigation?**

1) You will need to take one bolt, wind the wire on all the threads of the screw, and take the insulation off the wire's ends. Dip this into the container of BBs and pull it out. Did any BBs come out?

2) You will need to take the other bolt, wind the wire on half of the threads of the screw, and take the insulation off the wire's ends. Dip this into the container of BBs and pull it out. Did any BBs come out?

3) The bolts with the wire will be the electromagnets. Be careful when you do the next part of the experiment. The setup will get hot fast when you attach the wires to the battery. So you must do the next procedure calmly and quickly, taking a clip off the wire once you are done.
4) Place the half-wrapped bolt into the container with BBs. Attach the battery pack to both ends of the electromagnet wire and gently pull the magnet out of the container with the BBs, trying not to knock any BBs off. Place the electromagnet with the BBs over the empty container and detach one of the electromagnet wires, breaking the circuit.
5) Count how many BBs are in the new container. Give this data to the teacher when called on.

    # of BBs _____ half wrapped electromagnet.
6) Now place the fully wrapped bolt into the container with BBs. Attach the battery pack to both ends of the electromagnet's wires and gently pull the magnet out of the container with the BBs, trying not to knock any BBs off. Place the electromagnet with the BBs over the empty container and detach one of the electromagnet wires, breaking the circuit.
7) Count how many BBs are in the new container. Give this data to the teacher when called on.

    # of BBs _____ fully wrapped electromagnet.

8) Fill in Data Table 1 below as you find out how many BBs each group got for their data in the class.
9) Add the BBs in each column and divide by the number of groups; this will give the average number of BBs for both electromagnets.

Data Table 1

Class Group	# BBs on Half Wrapped Electromagnet	# of BBs on Fully Wrapped Electromagnet
1		
2		
3		
4		
5		
6		
7		
8		
9		
10		
11		
12		
13		
14		
15		
Average # of BBs		

Questions:

1) Did the bolts with the wire have any magnetic properties by themselves? How did you know?

2) Which electromagnet held the most BBs?

3) How is the magnetic force related to the number of turns on the wire?

4) How does wrapping the wire around the bolt increase the force of the electromagnet?

5) How is the number of BBs related to the magnetic force?

6) Why must the bolts in this experiment be identical?

7) How can electromagnets be more flexible than permanent magnets?

8) What do you think would happen if we use a stronger battery?

9) Have the teacher demo this. How many BBs were picked up this time?

10) How does voltage affect the force of the electromagnet?

Identifying Conductors and Insulators

Directions:

You will need a **battery**, a **battery pack** with wires exposed at the end, a **Christmas light** cut, and insulation stripped off the wires' ends. Put the battery(s) into your battery pack. Attach one end of the exposed wire of the battery pack to one exposed end of the Christmas light by twisting them together. This light will be used to see if the materials are conductors or insulators. Make sure it works by taking the free, exposed ends of the battery pack wire and light bulb wire and make them touch. If the bulb lights, it works.

1) You will also need a variety of materials like a **penny**, a **wooden spoon**, a **metal spoon** or **fork**, a **paper clip**, **paper**, a **comb**, **aluminum foil**, an **aluminum can** (check both the top of the can and the painted label), a **rubber band**, a **pencil**, and the **pencil lead** of a mechanical pencil (really carbon-graphite). **Looking at the materials and lab we will be using, what are the safety precautions we should take to protect ourselves and materials during the investigation?**

2) Test each of the materials you gathered in #1 by taking the exposed free ends of the wires from the battery pack and the Christmas light and touching them both on the material you are testing at the same time on different ends of the material. If the light bulb lights, it is a conductor because electrons can pass through the material. If it does not light, it is an insulator because it does not let electrons pass through the material. Fill in Data Table 1 below, listing conductors and insulators.

Data Table 1

Light bulb lights: Conductors	Light bulb does not light: Insulators

Questions:

1) What pattern do you see in the materials that are conductors?

2) What pattern do you see in the materials that are insulators?

3) What prevents the light bulb from lighting in insulators?

4) What other materials could allow the light bulb to light?

5) What other materials might cause the light bulb not to light?

6) Was there anything that lit the light bulb that surprised you? Discuss this with your teacher.

7) Explain how energy is used to light the light bulb?

Simple Middle School Physical Science Investigations   Seven Sides Publishing

Battery Power

Directions and Questions:

You will need two **batteries** and a string of three **Christmas lights in series** with the ends of the wires exposed. **Looking at the materials and lab we will be using, what are the safety precautions we should take to protect ourselves and materials during the investigation?**

1) What kind of energy does a battery contain?

2) Make a circuit lighting the bulbs by putting one wire on the battery's positive end (+) and the other wire on the battery's negative end (−). Notice how bright the lights are.

3) Do the same thing with two batteries stacked together. Which was brighter, one battery or two? Why?

4) What is the voltage difference (potential difference) of each battery?

5) Add them together and find the voltage difference (potential difference) for both together.

6) If a brighter light means a bigger current, what is the relationship between voltage difference (potential difference) and current?

7) What do you think will happen if you stack three batteries together?

8) How does the energy change as it transfers from the battery to produce light in the light bulb?

Making a Graphite Light Bulb (B)

Directions:

You will need **.2 to .5 mm graphite mechanical pencil lead** of different sizes, a **glass jar** with a **lid**, two **wires with alligator clips**, two **6-volt lantern batteries** (total of 12 volts), and **blue tac** (used to fix papers and posters to walls). When you cause resistance in a circuit, heat and light can be given off because the flow of electrons is forced to slow down. **Looking at the materials and lab we will be using, what are the safety precautions we should take to protect ourselves and materials during the investigation?**

1) Take two pieces of the blue tac and fix them to the inside lid of the jar. Place an alligator clip from both wires to the blue tac, holding the mouths up.
2) Take the smallest mechanical pencil lead (graphite) and break a piece off big enough to fit in the two alligator clips and inside the jar's lid. Place the glass jar over the top of the lid covering the graphite and alligator clips.
3) Turn off the lights in the room. Take the other ends of the wires and clip one wire to the "-" end of the first battery and another wire clipped to the "+" end of the second battery. To complete the circuit, clip one end of another wire on the "+" end of the first battery and clip the other end to the "–" end of the second battery. Watch it light up and notice how bright it is.
4) Make sure to disconnect the clips from the battery setup when you have finished.
5) Repeat the procedures for #s 2-4 for the different size graphite you have. Compare the brightness of each.

Questions:

1) Which size graphite lit up the brightest?

2) Which size graphite lit up the least?

3) Which graphite caused the most resistance to the flow of electrons? How do you know?

4) Which two variables caused the resistance in this investigation?

5) Why do you think the graphite lit up?

6) Look at the periodic table and find Carbon; this makes graphite. Why do you think a nonmetal was able to be used here?

7) Is Carbon a conductor or insulator?

8) When you slow the flow of electrons in a part of the circuit, what do you see given off?

9) How does the energy change as an incandescent light bulb lights up when a current flows through it?

10) Explain how electrons and photons are involved with what is happening inside this light bulb.

11) Why did we cover the graphite with the glass jar?

Human Circuits

Directions:

You will need a **Current Conductor** (found at Hobby Lobby) to show a circuit is closed by lighting up and making noise. **Looking at the materials and lab we will be using, what are the safety precautions we should take to protect ourselves and materials during the investigation?**

Series Circuit:

1) Have students make a circle and hold hands while two of them hold the Current Conductor between them on the metal strips. When everyone is touching hands in the circle, the circuit will be closed, and the Current Conductor will light up and make noise.

2) **Open the Circuit:** Have someone anywhere in the circle let go of someone's hand. What happened to the current of electricity?

3) Have everyone hold hands again, and at a different place, have someone let go. What happened?

4) Do you think this will happen every time in a series circuit?

Parallel Circuit:

5) Have a few students come out of the circle and make one or two lines (depending on how big your class is) connecting across the inside of the circle. Make sure everyone is still touching hands. How do we know the circuit is closed?

6) Have students on the circle opposite to the Current Conductor let go. What happened to the light and sound?

a. Why do you think this happened?

7) Where do people have to let go to make the circuit open? Experiment and find out.

8) If you had appliances in your house hooked up in a series circuit, what would happen if one of them was turned off?

9) How is a parallel circuit different than a series circuit?

10) Do you think your house is wired in series or parallel? Explain why.

11) How does this investigation show how energy is being transferred?

Simple Middle School Physical Science Investigations   Seven Sides Publishing

Comparing Series and Parallel Circuits

Directions and Questions:

You will need a **multi-meter**, **Christmas lights** in series three with **one bulb**, one with **two bulbs**, one with **three bulbs**, a **battery pack,** and two **batteries. Looking at the materials and lab we will be using, what are the safety precautions we should take to protect ourselves and materials during the investigation?**

Hypotheses

1) Light bulbs are resisters. They resist or slow the flow of electrons in a circuit, which causes the light bulbs to glow. Predict which lights will glow brighter: three bulbs in series or three bulbs in parallel circuits.

2) Predict which set of lights will stay lit when a bulb is removed: bulbs in series or bulbs in parallel.

Experiment

3) Compare the lights' brightness when the lights are put into a closed series circuit with 1, 2, and 3 bulbs.

4) Which group of bulbs was the brightest?

5) Which group was the dimmest?

6) Why do you think it happened that way?

7) Compare the brightness of lights when the lights are put into a parallel circuit with 1, 2, and 3 bulbs.

8) Is this the same or different from the series circuit? If different, explain how it is different.

9) Take the three bulbs in series, remove one light carefully, and connect to the battery pack. Do they light up?

10) Take the three bulbs in a parallel circuit, remove one light carefully, and connect it to the battery pack. Do they light up?

11) Why do you think the results from #s 9 & 10 came out the way they did?

12) How did your results compare to your hypotheses?

13) How does the energy flow through a series circuit different from a parallel circuit?

14) What is causing that difference?

Simple Middle School Physical Science Investigations   Seven Sides Publishing

Measuring Actual Voltage and Resistance

Directions and Questions:

You will need a **multi-meter**, **Christmas lights** in series three with **one bulb**, one with **two bulbs**, one with **three bulbs**, a **battery pack,** and two **batteries. Looking at the materials and lab we will be using, what are the safety precautions we should take to protect ourselves and materials during the investigation?**

Checking Voltage

1) Take a battery out of the battery pack and measure the battery's voltage.

2) Now put the battery back and measure the voltage going through the wires.

3) How is the answer in #2 related to your answer to #1?

4) **Hypothesis.** How do you think voltage will change as you add bulbs in a series circuit?

5) Connect one light to the battery pack and measure the voltage going through one light?

6) Now take off the one light, put on the two lights in series, and measure the voltage going across the lights.

7) Now check the voltage going across three lights in series.

8) **Conclusion.** How are #s 5, 6, and 7 related?

9) **Hypothesis.** How do you think the voltage will change as you add bulbs in a parallel circuit?

10) Check the voltage going across two bulbs in a parallel circuit.

11) Predict what voltage will go across three bulbs in parallel.

12) Now check the voltage going through 3 bulbs in a parallel circuit.

13) **Conclusion.** What do you think you can say about the voltage going through a circuit?

Checking Resistance

14) **Hypothesis.** How do you think the resistance will change as we add bulbs in the series?

15) Check the resistance of one bulb.

16) Check the resistance of 2 bulbs in series.

17) Check the resistance of 3 bulbs in series.

18) **Conclusion.** How does resistance change as you add bulbs in series?

19) **Hypothesis.** How do you think the resistance will change as we add bulbs in parallel?

20) Check the resistance of 2 bulbs in a parallel circuit.

21) Check the resistance of 3 bulbs in a parallel circuit.

22) **Conclusion.** How does the resistance change as you add bulbs in a parallel circuit?

23) How does the energy flow differently through series and parallel circuits?

Hotdog Circuits

Warnings and Directions:

Follow the teacher's directions carefully!

Once you plug in the **extension cord with large alligator clips** attached to each wire, do not touch the **forks** or the **hotdogs**, or you will get shocked. You will also need other **cords with large alligator clips** on both ends but no plug and a **pickle. Looking at the materials and lab we will be using, what are the safety precautions we should take to protect ourselves and materials during the investigation?**

1) Follow the directions to make the systems below, and once plugged in, time how long it takes to cook each hotdog. Write this in Data Table 1. Turn on the stopwatch when you plug in the cord and turn it off when you see the hotdog skin bubble or split. You can then cook longer to your desired taste. **Once your hotdog is done, unplug the cord at the plug before touching anything.**

2) With what you know about circuits, which do you think will cook faster, hotdogs in series or parallel circuits?

3) Have one set-up where you will have one hotdog between two forks with an alligator clip on each fork from the extension cord. Do not plug in the cord until you have set the forks in each end of the hotdogs and the clips are on the forks. **Touching any part of the apparatus (including the hotdog) could cause a damaging shock. Do not touch it until you unplug the set-up. Check with your teacher before you start if you have any questions.**

4) Make another set-up with two hotdogs and three hotdogs each in series. You will need to have pieces of cord with no plugs, but alligator clips on each end of the forks put in the hotdogs to connect between the forks of each hotdog so the circuit can run through all the hotdogs. Use the big extension cord with the plug to connect to the hot dogs on each end. **Remember not to plug the cord into the electrical outlet until everything is set up. Once it is plugged in, do not touch it until you unplug each set-up. Check with your teacher before you start if you have any questions.**

5) Make another set-up with two hotdogs in parallel and three in parallel. You do this by poking the forks into each end of each hotdog, then stacking or overlapping the handle ends of the forks over each other, placing them inside each alligator clip of the cord with a plug. **Remember not to plug the cord into the electrical outlet until everything is set up. Once it is plugged in, please do not touch it until you unplug each set-up. Check with your teacher before you start if you have any questions.**

Data Table 1

Number of Hotdogs	Time cooked in series	Time cooked in parallel
1		
2		
3		

Questions:

1) How long did it take to cook one hotdog between two forks?

2) How long did it take for the two hotdogs to cook in series?

3) How long did it take for the two hotdogs to cook in a parallel circuit?

4) Write down how long it took to cook three hotdogs in series and parallel circuits:

   3 in series: _____ 3 in parallel: _____

5) What part of a circuit does the hotdog represent?

6) What causes the hotdog to cook?

7) If you needed to cook as many hotdogs as you can, as fast as possible, using an electric circuit, which type of circuit would you use and why?

8) Compare and contrast how electricity flows through series and parallel circuits.

9) How does the energy change as the electricity heats the hotdog?

Pickle Directions and Questions:

10) Set up a pickle between two forks like you did with the hotdogs. Turn off the lights in the room, and then plug in the cord. Do not touch any part of the apparatus while the pickle is plugged in. What happens when we create a circuit with a **pickle** instead of a hotdog?

11) A pickle is soaked in vinegar (acidic acid); how do you think this affected the electrons in the pickle when plugged in?

12) Do you think what happened with the pickle will happen with a fresh cucumber? Explain why.

13) Explain how the energy changes as electricity is used to light up the pickle.

Motors and Generators

Directions and Questions:

You will need various strength **batteries, scrap paper, scissors,** a **simple motor**, and a **flashlight generator**. **Looking at the lab and materials we will be using, what are the safety precautions we should take to protect ourselves and materials during the investigation?**

Motors

1) With your scissors, cut out a simple propeller that you will push over the motor's axle to spin. Take the smallest voltage battery and touch the wires to the positive and negative terminals of the battery. Which direction did the propeller spin?

2) Switch the wires to touch the opposite terminals of the battery. What direction did the propeller spin now?

3) Notice how fast the propeller spins with this battery. Take the next higher voltage battery and attach the wires to the battery's positive and negative terminals. How fast does the propeller spin now compared to the last battery?

4) How do you think the propeller will spin with an even higher voltage battery?

5) How is energy changed as electricity moves through the motor?

6) Describe how you can control a motor's direction and speed. Discuss with your teacher.

Generators

1) Take your flashlight generator and press the handle to make the magnet spin inside the copper wires wrapped around an iron core. What happens to the light bulb as you do this?

2) Speed up how fast you move the magnet in the copper wires. What happens to the light bulb?

3) How is electrical energy produced in a generator? Discuss with your teacher.

4) How else could you move a magnet in copper wires wrapped in an iron core or move copper wires wrapped around an iron core around a magnet?

Comparing Motors and Generators

1) A motor uses a permanent magnet and copper wires around an iron core with a current (which makes an electromagnet) to spin the motor's axle. How is this similar to the generator?

2) Compare and contrast how the energy is transferred through an electric motor and a generator.

3) Could a motor be used as a generator? If so, explain how.

Virtual Investigations that go with Electromagnetism

ExploreLearning.com:

    Coulomb Force (Static) Gizmo

    Charge Launcher Gizmo

    Pith Ball Lab Gizmo

    Electrostatic Induction Gizmo

    Polarity and Intermolecular Forces Gizmo

    Circuit Builder Gizmo

    Circuits Gizmo

    Advanced Circuits Gizmo

    Magnetism Gizmo

    Magnetic Induction Gizmo

    Electromagnetic Induction Gizmo

PhET.colorado.edu:

    Balloons and Static Electricity

    Capacitor Lab

    Capacitor Lab Basics

    Charges and Fields

    Conductivity

    Coulomb's Law

    Electric Field Hockey

    Electric Field of Dreams

    John Travoltage

    Radiating Charge

- Static
- Battery Voltage
- Battery-Resistor Circuit
- Circuit Construction Kit (AC+DC)
- Circuit Construction Kit (AC+DC) Virtual Lab
- Circuit Construction Kit DC
- Circuit Construction Kit DC-Virtual
- Conductivity
- Ohm's Law
- Resistance in a Wire
- Semiconductors
- Single Circuit
- Charges and Fields
- Electric Fields of Dreams
- Electric Field Hockey
- Faraday's Electromagnetic Lab
- Faraday's Law
- Generator
- Magnet and Compass
- Magnets and Electromagnets
- Radiating Charge

Physicsclassroom.com:

Physics Interactives:

Static Electricity

Aluminum Can Polarization

Charging

Name that Charge

Coulomb's Law

Electric Field Lines

Put the Charge in the Goal

Electrostatics Landscapes

Magnetic Fields

Electric Circuits

DC Circuit Builder

Concept Builders:

Static Electricity

Charge and Charging

Charge Interactions

Triboelectric Charging

Polarization

Charging by Induction

Coulomb's Law

Electric Field Intensity

Electric Circuits

Light Bulb Anatomy

Current

Resistance Ranking Tasks

Know Your Potential

I =ΔV/R Equations as a Guide to Thinking

Which One Doesn't Belong? – Equivalent Resistance

Series Circuits – ΔV = I x R Calculations

Parallel Circuits – ΔV + I x R Calculations

Unit 11: Waves

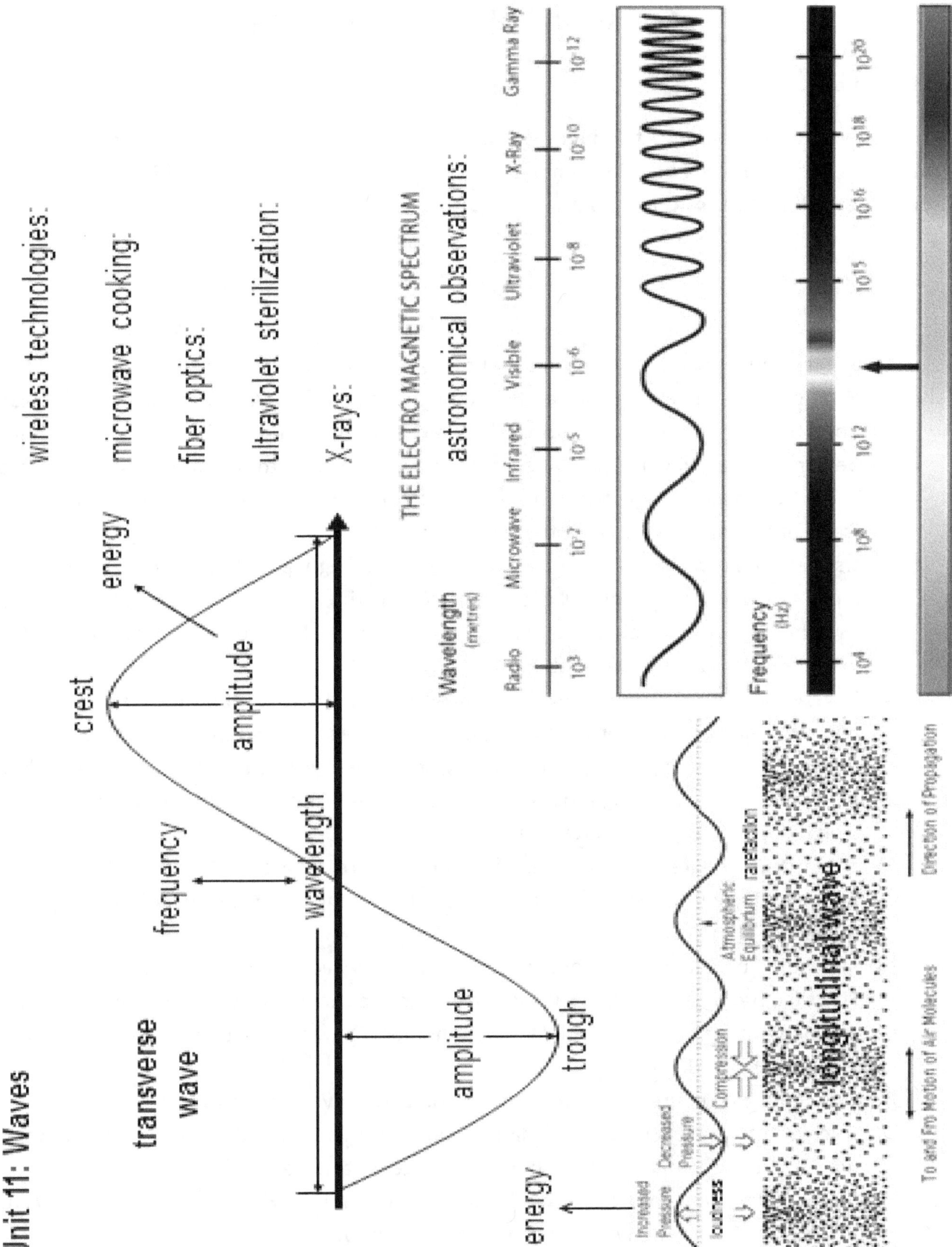

Measuring Wave Properties

Directions:

You will need a long springy **telephone cord**, a **stopwatch**, and a **meter stick**. **Looking at the materials and lab we will be using, what are the safety precautions we should take to protect ourselves and materials during the investigation?**

1) Stretch your phone cord and measure how long it is when you will make waves with it.
2) Have two people, one on each end, hold the phone cord. Have one person send a pulse down the cord by having them move their hand back and forth once. Have a third person measure how long it takes for the wave to make it to the other person. In Data Table 1 on the next page, write this data for all three wave times (since all the waves will travel at the same speed down the same medium).
3) Oscillate the cord back and forth to create a standing wave that is half a wavelength long; this will make it look like one hump is in the wave. Make sure you keep the rhythm. In your data table, multiply your cord's length by 2 to get the wavelength for wave one. Write this in Data Table 1.
4) Have the person with the stopwatch; once they see the rhythm on the cord, say start, and they will start the stopwatch and measure 10 seconds and then say stop when the ten seconds are done. The person moving their hand will count how many times they move their hand to the right during the 10 seconds. Write this data down in Data Table 1 on the next page as the wave count.
5) This time, oscillate the cord back and forth to create a standing wave that is one wavelength long; this will look like there are two humps. The wavelength for wave 2 is the length of the cord.
6) Repeat the procedure for #4 and write the data for wave 2 in Data Table 1.
7) This time oscillate the cord back and forth to create a standing wave with one and a half wavelengths long; this will look like three humps. The wavelength for wave 3 is 2/3 the length of the cord.
8) Repeat the procedure for #4 and write the data for wave 3 in Data Table 1.
9) Calculate the wave speed by taking the length of the cord and dividing it by the wave time. Write this in Data Table 1 on the next page for all three waves.
10) Calculate the frequency by taking the wave count and dividing it by 10 seconds. Write this in Data Table 1 on the next page for all three waves.

Data Table 1

Wave	Cord Length (m)	Wave Time (s)	Wave Speed (m/s)	Wave Count	Wavelength (m)	Frequency (Hz)
1						
2						
3						

Questions:

1) Why was the wave speed the same for all three waves?

2) What is the relationship between wavelength and frequency?

3) The cord has a natural frequency and what you created were three octaves of that frequency. Where is the word octaves used that has to do with waves?

4) How does frequency relate to pitch?

5) So if someone sings at a high pitch, is the wavelength long or short? Explain why.

6) If someone sings at a low pitch, is the wavelength long or short? Explain why.

Observing Waves in a Slinky

Directions:

You will need a standard to long **slinky** to make both compression waves and transverse waves. You will do this with one person holding one end of the slinky and another person holding the other. One person will move one end; the other will hold still. Each person needs to take a turn moving the slinky to make the waves described below. You will show this to your teacher. **Looking at the material and lab we will be using, what are the safety precautions we should take to protect ourselves and materials during the investigation?**

1) Make a compression wave with high frequency by moving the end of your slinky forward and backward in the same plane as the slinky quickly.

2) Make a compression wave with low frequency by moving the end of the slinky forward and backward in the same plane as the slinky but slowly.

3) Make a compression wave with high amplitude by repeating the procedure in #2 but making bigger, more violent pushes down the slinky.

4) Make a compression wave with low amplitude, the same as the procedure in #2 but making smaller, less violent pushes down the slinky.

5) Make a transverse wave with high frequency by quickly moving the slinky's end perpendicular to the slinky.

6) Make a low-frequency transverse wave by repeating the procedure in #5 but move the slinky more slowly.

7) Make a transverse wave with a high amplitude by repeating the procedure in #5 but moving the end of the slinky a bigger distance perpendicular to the slinky.

8) Make a transverse wave with low amplitude by repeating the procedure in #5 but not as big a distance as #7.

9) Now make a compression wave with high frequency and low amplitude.

10) Now make a compression wave with low frequency and high amplitude.

11) Now make a transverse wave with high frequency and high amplitude.

12) Now make a transverse wave with low frequency and low amplitude.

Observing Sound

Directions:

You will need a **wire hanger** and **string**. Tie two pieces of string on either side of the wire hanger. **Looking at the materials and lab we will be using, what are the safety precautions we should take to protect ourselves and materials during the investigation?**

1) Wrap the string around each of your index fingers and clank the hanger against your desk. How does the hanger sound?

2) Put your fingers in your ears and clank the wire hanger against the desk. How does the hanger sound now?

Questions:

1) Do sounds travel better through the air or the string?

2) Which medium do you think sound travels the fastest through? Put the types of medium (the solid, liquid, and gas) in order from fastest to slowest.

3) What if there is no medium for sound to travel through, will there be any sound?

4) When the Death Star blows up in Star Wars, can that really make a sound in outer space? Explain why?

Simple Middle School Physical Science Investigations   Seven Sides Publishing

Coffee Can Phones

Directions and Questions:

Take two **coffee cans** and poke a small hole in the bottom of each of them. Cut a long piece of **string** that reaches across the room, put the ends through each can, and tie a knot in them, fixing them to both cans. **Looking at the materials and lab we will be using, what are the safety precautions we should take to protect ourselves and materials during the investigation?**

1) Pull the string tight while holding the cans and talk through them. Can you be heard in the other can? Why do you think that is?

2) Have someone pinch the string with their fingers halfway across. Can you be heard in the other can now? Why do you think that is?

3) Let the string loosen and droop. Talk again. Can you be heard in the other can? Why do you think that is?

4) Combine with a group next to you, cross your strings, and have them touch while the strings are taught. Have someone talk into a can. Who can hear in their cans?

5) Have someone pinch with their fingers where the strings cross. Can anyone hear now?

6) Explain how sound travels from one can to the other.

Music Test Tubes

Directions:

You will need four **test tubes**, a **test tube holder**, and different **water** amounts placed in each test tube. Test tube one, leave empty. Test tube 2 fill it ¼ full of water. Test tube 3 fill 1/3 with water. Test tube 4 fill with ½ full with water. **Looking at the materials and lab we will be using, what are the safety precautions we should take to protect ourselves and materials during the investigation?**

1) Predict how the toned will sound different when you blow across the top of the test tube. Rank the predicted sounds from 4 being the lowest to 1 being the highest. Write that in Data Table 1 below.
2) Blow across test tube one until you get a tone produced. Do the same for each of the test tubes. Rank the order from the 4 being the lowest tone to 1 being the highest.
3) The tone is created by how large/long the tube is for the air to vibrate. The longer the tube, the longer the wavelength. Empty and wash the test tubes when you are done.

Data Table 1

Test Tube	Amount of Water	Predicted Tone Differences	Tone Difference
1	empty		
2	¼ full		
3	1/3 full		
4	½ full		

Questions:

1) Describe how the tones changed depending on the amount of water in the test tube.

2) How did the pitch depend on the height of the water?

3) Why are the tones different from the different test tubes?

4) Explain how resonance amplifies the sound of a test tube.

5) How do the natural frequencies of the columns of air in each test tube differ?

6) Compare how the test tubes make music with how a flute makes music.

7) How is the flute different from the test tubes?

8) How does the music industry use the physics of waves?

Singing Glasses and the Dancing Toothpick

Directions and Questions Part 1:

You will need a **flat toothpick, crystal glasses of different sizes**, some **crystal glasses of the same size** filled with different **water** amounts, and the **internet. Looking at the materials we will be using, what are the safety precautions we should take to protect ourselves and materials during this investigation?**

1) Wet your finger and rub it around the rim of one of the glasses until you hear a hum.
2) Repeat the procedure in #1 for different sizes of glasses. Which glasses (larger or smaller) have a deeper tone or pitch?

3) Which glasses have a higher tone or pitch?

4) Fill some of the glasses of the same size with different amounts of water. Which glass has the lowest pitch?

5) Which glass has the highest pitch?

6) Research on the internet why your results came out the way they did. What causes the sounds to change when you change the glass's size or how much water is in the glass?

Directions and Questions Part 2:

1) Now, take two of the glasses you used in Part 1 that are the same size and place them near each other. Balance a flat toothpick on the rim of one of the glasses.
2) Rub the rim of the other glass to make it hum. What do you notice happens to the toothpick?

3) Discuss with your class and teacher, then explain why this happens.

4) What do natural frequency and resonance have to do with this phenomenon?

Playing the Rubber Band

Directions:

You will need a **plastic tub** and three different **rubber bands** of the same length with different widths. **Looking at the materials and lab we will be using, what are the safety precautions we should take to protect ourselves and materials during the investigation?**

1) Find the mass of each of the rubber bands. Write that in Data Table 1 below.
2) Put the rubber bands around the tub. Try to have the same tension on each rubber band.
3) Pluck each rubber band and tell how they differ. Rank the pitch from 3 being the lowest to 1 being the highest. Put that in Data Table 1 below.

Data Table 1

Rubber Band	Mass	Pitch
Skinny		
Medium		
Thick		

Questions:

1) How does the width of the rubber band affect the pitch?

2) The rubber band that takes the longest to move back and forth will have the lowest frequency. Try to explain why this happens. (Remember the Law of Inertia.)

3) How do you think that affects the acceleration of the rubber bands as they move back and forth?

4) How do you think the length of the rubber band would affect the pitch?

5) How do string thickness and length affect how string instruments sound and are played?

Music has Patterns

Part 1 Directions:

You will need a **digital keyboard** and a **microphone probe** attached to an **interface** connected to a **computer** with **Logger Pro. Looking at the materials and lab we will be using, what are the safety precautions we should take to protect ourselves and materials during the investigation?**

1) In Logger Pro, open the Physics with Vernier folder and file #35, Mathematics of Music.
2) Position the microphone near the opening where the sound comes out of the instrument. Press "Collect." Play a middle C for the first note. Play it until you see the wave form on the screen. Record the frequency in Hz. Write this in Data Table 1.
3) Now play the next higher note to repeat the procedure in #2. Repeat this until you have recorded all the notes' frequency going up in the scale until you reach the next C.
4) Now calculate the ratio of the first note to middle C; this is done by taking the current note's frequency and dividing it by the middle C frequency.
5) Take the decimal from the ratio and try to find the fraction that is closest to that decimal, and write your answer like the example shown in Data Table 1. Compare this to the different wrench sizes we use in a toolbox or on a ruler in proportions to an inch; you will see a pattern. Any variation off of the pattern shows the instrument is out of tune.

Data Table 1

Key	Note	Frequency (Hz)	Ratio to C	Ratio to C Fraction
1	C			1 and 0
2	D			
3	E			1 & ¼
4	F			
5	G			

6	A			1 & 2/3
7	B			
8	C			

Questions Part 1:

1) What is the frequency ratio to middle C with the next C higher?

2) How long is the wavelength with this C compared to the middle C? Hint: use the wave formula.

3) Are there any other notes that we did not play? What do you think they are?

Part 2 Directions:

1) To see why they are there, play all the keys in order from middle C (even the black keys). Write the frequencies in Hz down in Data Table 2.
2) Then, find the ratio to the previous note by taking the current note and dividing it by the frequency of the previous one. What number did you get for each?

Data Table 2

Key	Note	Frequency	Ratio to the prev. note
1 White	C		
2 Black	C sharp		
3 White	D		
4 Black	E flat		
5 White	E		
6 White	F		

7 Black	F sharp		
8 White	G		
9 Black	A flat		
10 White	A		
11 Black	B flat		
12 White	B		
13 White	C		

Questions Part 2:

1) What number did you get for each ratio to the previous note?

2) Besides the pattern of frequency ratios, what other patterns do we see in music?

3) If there are no patterns in the sound, what do we call it?

4) What should you have in sound to be able to call it music?

The Doppler Effect

Directions and Questions:

You will need a toy **football that whistles** when you throw it. Stand between two people that will throw it back and forth over your head. **Looking at the materials and lab we will be using, what are the safety precautions we should take to protect ourselves and materials during the investigation?**

1) When the ball is flying over your head, pay attention to the whistle's pitch. How was the sound before the ball reached you compared to the sound after it passed you?

2) What happens to the pitch of the whistle as it passes over your head?

3) Why do you think this happened?

4) When would the ball move into the waves, shortening the wavelength coming to your ear as it flies in the air?

5) When would the ball move away from the waves, lengthening the wavelength coming into your ear as it flies in the air?

6) How does this explain why a siren sounds higher as it approaches and changes to a lower pitch as it passes?

Simple Middle School Physical Science Investigations   Seven Sides Publishing

Making a Rainbow

Directions and Questions:

You will need a **sunny day** and a **hose** or **spray bottle** to make mist. **Looking at the materials and lab we will be using, what are the safety precautions we should take to protect ourselves and materials during the investigation?**

1) Make mist with a hose or a spray bottle with the sun behind you until you see a rainbow. Try to see the entire rainbow. What is the shape of the rainbow?

2) What is the order of the colors of the rainbow?

3) Which color is on the outside?

4) Which color is on the inside?

5) Which color do you think has the longest wavelength? Explain why.

6) Which color do you think has the shortest wavelength? Explain why.

7) AM radio can be heard across the country because the wavelength is longer and can get past the curvature of the Earth; this is not the case for FM radio; once out of town, the radio station seems to go out. Which color do you think would be able to go around the particles in the air when the light passes through the thickest part of the atmosphere? Explain why.

8) What color is the sun when we see it on the horizon (sunrise or sunset)? Explain why.

9) What color is the moon when we see it on the horizon? Explain why.

Extention: If you have some **diffraction gradient glasses**, put those on and explain what you see happen to all the light you see. These act similar to how the water diffracts the light we just saw through the rainbow.

Polarization of Light

Directions:

You will need a **light source**, two **polarizing filters**, and a **light sensor** attached to an **interface** connected to a **computer** with **Logger Pro. Looking at the materials and lab we will be using, what are the safety precautions we should take to protect ourselves and materials during the investigation?**

1) In Logger Pro, open the folder Physics with Vernier and the file #28 Polarization of Light. Press "Collect."
2) Turn on your light source and turn off all the other lights in the room. Place your light sensor at a distance away from your light source and measure the light intensity. Write the data in Data Table 1 below.
3) Place a polarizing filter between the sensor and the light. Measure the light intensity and write the data in Data Table 1 below.
4) Place another polarizing filter between the sensor and the light in a way that still allows light to pass through. Measure the light intensity and write the data in Data Table 1 below.
5) Rotate one filter 90° to block the light from the light source going to the sensor. Measure the light intensity now and write that data in Data Table 1 below.

Data Table 1

Setup	Light Intensity
Just light	
Light through a filter	
Light through 2 filters	
Light through 2 filters at 90°	

Questions:

1) How did the light intensity change when the filter was placed in front of the light?

2) How did the light intensity change when two filters were placed in front of the light?

3) What happened to the light intensity when one lens was rotated 90°?

4) Hold the two filters up to the light and rotate one. What do you see?

5) Why do you think this happens?

6) Polarized sunglasses do the same thing. How can you tell if a pair of sunglasses are really polarized?

3D Glasses

Directions and Questions:

You will need two pairs of **3D glasses. Looking at the materials and lab we will be using, what are the safety precautions we should take to protect ourselves and materials during the investigation?**

1) Face the glasses towards each other directly in front of each other. What color do you see through the lenses when looking through both sets of lenses simultaneously?

2) Move the glasses closest to you back and forth from left to right. How do the colors change when looking through both sets of lenses on the glasses simultaneously?

3) Turn one set of glasses perpendicular to the ground. How did the color change as you looked through both sets of lenses?

4) Face the glasses in the same direction (looking at them from behind) and turn one pair perpendicular to the ground; how does the lens color change when looking through both sets of lenses of the glasses at the same time?

5) Put one pair of glasses on and look at the other with both eyes open. How do the lenses look on the other pair of glasses?

6) Close your right eye; what do you see now in the other glasses?

7) Close your left eye; what do you see now in the other glasses?

8) Polarized sunglasses block the horizontal light from hitting your eyes and allow the vertical light through; this keeps you from seeing glare from the sun reflecting off surfaces. How could you tell if lenses at the sunglasses store are really polarized?

9) The slits on the 3D glasses have one lens with microscopic slits that block light coming at you vertically, whereas the other lens blocks the light coming at you horizontally. If they are turned at right angles to each other, it blocks all light. Two projectors show the same movie, just staggered with different types of light so that your right eye sees one projector, and your left eye sees the other; this allows you to see a 3D image on the screen.

Light Pipes

Directions and Questions:

You will need **fiber optics** to observe some **light sources of different colors. Looking at the materials and lab we will be using, what are the safety precautions we should take to protect ourselves and materials during the investigation?**

1) Shine a light on the side of the fiber optics. Can you see the light coming out the ends?

2) Shine light into the end of a fiber optic wire. Do you see the light coming out the other end?

   a. Can you see the light coming out the sides?

3) Keep shining the light through one end of the fiber optic and bend the fiber optic. Can you still see the light coming out the other end?

   a. Can you see it coming out the sides?

4) Try shining the light on the other end. Does light still come out at the opposite end?

5) Why do you think they call these light pipes?

6) How can they be used to send information from one computer to another to have the computers talk to each other?

7) Who do you think benefits from this technology and information?

8) Try shining a different color of light down one end of the pipe. What do you see?

Simple Middle School Physical Science Investigations Seven Sides Publishing

Water Refraction

Directions:

You will need a **square tank** half-filled with **water**, an **Erlenmeyer flask** filled with **water**, and a **ruler. Looking at the materials and lab we will be using, what are the safety precautions we should take to protect ourselves and materials during the investigation?**

1) Put the ruler behind the tank with it up and down perpendicular to the floor. Look through the water from the front at the same level as the tank. Draw what you see below.

2) Look at the ruler through the tank from the front above the tank. Draw what you see below.

3) Turn the tank so that a corner is facing you. Put the ruler on the right-hand side; where do you see the ruler? Draw it below.

Questions:

1) Why do you think the images appear different through the same tank of water at different angles?

2) How would the image you see be different if the water's surface reflected light like a mirror instead of bending it?

3) How would the image change if the container was curved, bending out?

4) Pull out an Erlenmeyer flask filled with water. Look through it, and tell how things look different at different distances.

5) How does this explain how convex lenses work?

Test Tube Lenses

Directions:

Fill a **glass test tube** with **water** and seal it with a **rubber stopper**. Keep your finger or thumb over the rubber stopper so that the stopper does not fall off and spill water. **Looking at the materials and lab we will be using, what are the safety precautions we should take to protect ourselves and materials during the investigation?**

1) Set the test tube on the paper over the title of this lab. Write your observation in Data Table 1 about what you see through the test tube.
2) Hold the test tube approximately 1 cm over the title of this lab and observe it again. Record your observations in Data Table 1 below.
3) Repeat this three more times, increasing height approximately a centimeter each time. Write your observations in Data Table 1 below.

Data Table 1

Height	Observation of **Test Tube Lenses**
Right on the surface	
1 cm above the surface	
2 cm above the surface	
3 cm above the surface	
4 cm above the surface	

Questions:

1) Are the images you see real or virtual?

2) How high above the surface did the image become inverted?

3) What kind of lens does the test tube make (concave or convex)?

4) How does a magnifying glass compare to the test tube you looked through?

Reflection Lab

Directions:

You will need a **flat mirror, paper,** and a **protractor. Looking at the materials and lab we will be using, what are the safety precautions we should take to protect ourselves and materials during the investigation?**

1) Have your mirror facing you sitting on this paper where it says Place Mirror Here where three angles are drawn on it going into the mirror and a normal.
2) Ensure the normal is straight in and out of the mirror to properly see each of those lines pointing at you in the mirror; draw extensions for each line as if it came out of the mirror.
3) Measure the angle of incidence of the lines already there with respect to the normal.
4) Now measure the angles of reflection.

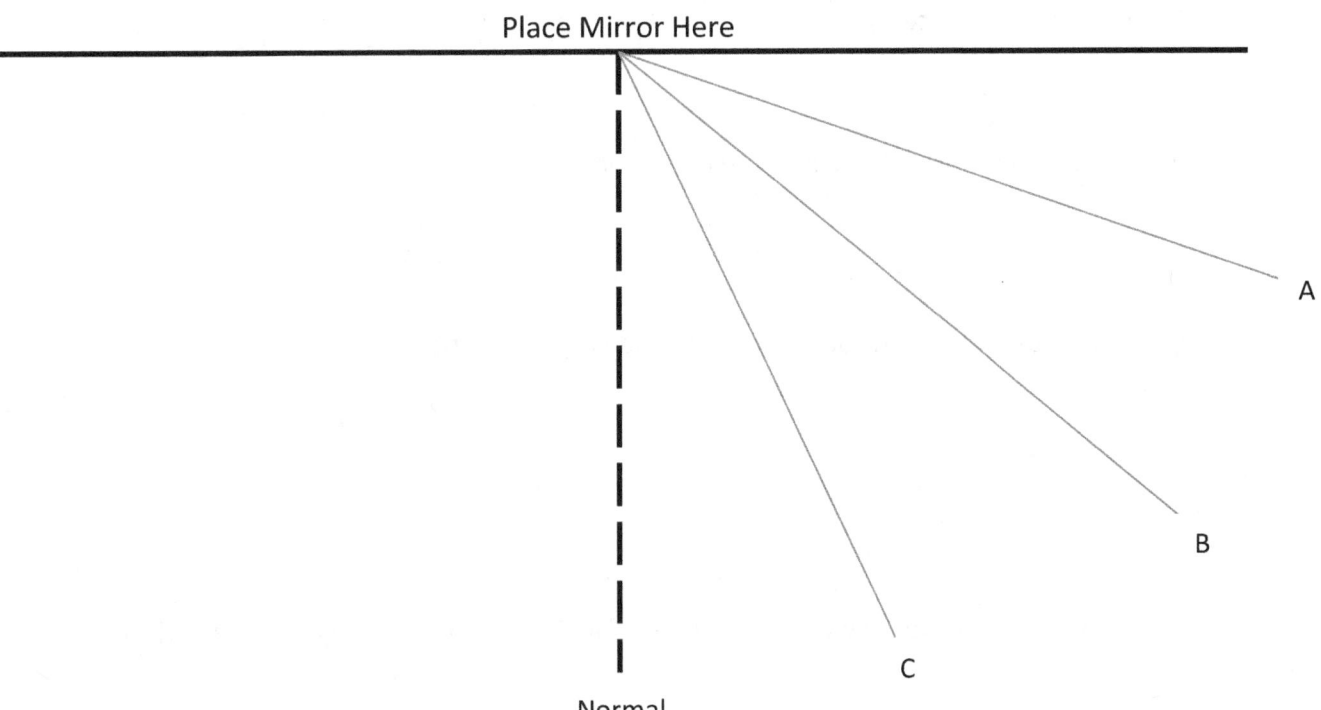

Questions:

1) What is the angle of incidence of ray A?

2) What is the angle of reflection of ray A?

3) What is the angle of incidence of ray B?

4) What is the angle of reflection of ray B?

5) What is the angle of incidence of ray C?

6) What is the angle of reflection of ray C?

7) What do you notice about the angles of the ray of reflections you drew compared to the angles of incidence?

8) Write a rule that states what you observed.

9) How would this information help you play a game of pool?

10) What is the line called where the angle of incidence and the angle of reflection is zero?

Magnifying Power

Directions:

You will need a **direct light source (bright lamp, flashlight, or sun)**, two different **power (thicknesses) lenses/magnifying glass**, and a **metric ruler**. **Looking at the materials and lab we will be using, what are the safety precautions we should take to protect ourselves and materials during the investigation?**

1) To find the focal length of both lenses, you will want to have the light source behind/above you and any flat surface in front of you. Move your lens forward and backward until you find the smallest diameter of a light spot on the front surface in front of you. Measure the distance your lens is from that surface; this is your focal length. Write this down in Data Table 1.
2) Calculate the magnifying power using focal length in cm with the equation:

   MP = 25cm/FL where MP = Magnifying Power, FL= Focal Length. Write this number in Data Table 1 for magnifying power using focal length.
3) To find the magnifying power using the length of an image you see, draw a 1 cm line on this paper, then place the lens over the image and see it clearly and larger. Hold your metric ruler up just above the image so you can see both the ruler and the line. Measure how long the line appears in the lens in centimeters; this now tells you the magnifying power. Place the magnifying power using the length of the image for both lenses in Data Table 1.

Data Table 1

Lens	Focal Length (cm)	Magnifying Power using Focal Length	Magnifying Power using Length of Image
Thick	(cm)		
Thin	(cm)		

Questions:

1) Describe the image formed by the magnifying glass/lens.

2) Is the image you see through the lens a real or virtual image? Tell why.

3) Compare the methods of calculating Magnifying Power. How close are the results?

4) Which method do you think is most accurate? Tell why.

5) What is happening to the light as it passes through the lens to cause the image to change?

6) How does this relate to the curvature of the lens?

7) In the movie *A Bug's Life*, an ant uses a drop of water as a lens. Is it possible to use a drop of water as a lens?

Simple Middle School Physical Science Investigations   Seven Sides Publishing

Brightness and Distance

Directions:

You will need a **light source,** a **laser light,** a **meter stick,** a **light sensor** attached to an **interface** connected to a **computer** with **Logger Pro,** or a **Vernier Dynamics System and Optics Expansion Kit. Looking at the materials and lab we will be using, what are the safety precautions we should take to protect ourselves and materials during the investigation?**

1) In Logger Pro, open the folder Physics with Vernier and file #29 Light Brightness Distance. Press "Collect."
2) Turn on your light source and turn off the other lights in the room. Place the 0 of the meter stick at the light source. Then place the front of the light sensor at each distance in Data Table 1 below. Measure the light intensity for each of those distances. Press "Keep" when the intensity value stabilizes at each distance. Write this data in Data Table 1 below, and press "Stop" when you have finished collecting all of the data.

Data Table 1

Distance (cm)	Intensity
5	
10	
15	
20	
25	
30	
35	
40	
45	
50	

Questions:

1) In Logger Pro, look at the graph of light intensity vs. distance. What is the relationship between distance and intensity?

2) Why do you think this happens?

3) How does this relationship help us find the distance between galaxies and stars?

4) How does this relationship relate to planets, and how close they can be to stars and possibly support life (the goldilocks zone)?

5) If the orbit of the Earth is not stable, when would temperatures be higher than normal?

   a. When would it be lower than normal?

6) Lasers are focused light made by mirrors and lenses. Try to shine a laser light into the light sensor. How does that compare to the other readings?

7) Change the distance the sensor is from the laser. Did the intensity change? Why do you think that is?

Uses of the Electromagnetic Spectrum

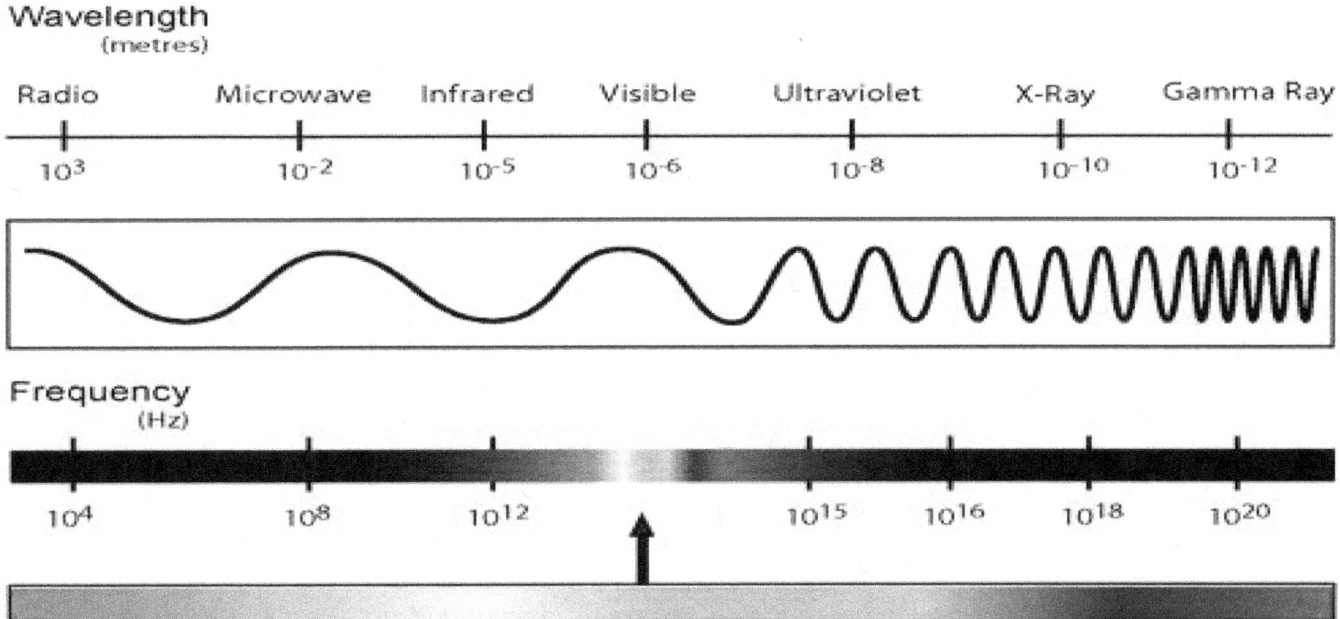

Directions and Questions:

Use the Electromagnetic Spectrum above and the **internet** to answer the questions below.

1) Electromagnetic waves have higher energy with shorter wavelengths and lower energy with longer wavelengths. Which waves above would have the highest energy, also being the most dangerous?

2) Which waves above have the lowest energy and are the most harmless?

3) Where is visible light on this spectrum?

4) Why do you think cell phones use microwaves to communicate?

5) Longer wavelengths can travel around corners easier. Which would travel farther, radio waves or microwaves?

6) Which would you see more of radio towers or cell phone towers? Explain why.

7) Why do you think remote controls use infrared waves to control TVs, drones, and remote control cars?

8) How do you think a pilot in Arizona controls a drone in the Middle East?

9) Why do you think the doctor puts a lead shield over you when you get an x-ray?

10) What do you think will give you a more detailed image, a DVD using red laser light or a DVD using blue? Explain why.

11) When using a telescope to look into outer space, which will give you more information and details, using the light waves or x-rays? Explain why.

12) Since the big Bang occurred 13.77 billion years ago, and its energy is lost over time, which category of waves would you expect to find its echo? Explain why.

13) Why do you think hospitals and restaurants use ultraviolet rays for sterilization?

14) Why do you think x-rays can see our bones?

How we use Microwave Ovens

Directions:

You will need a **microwave oven**, two **microwave bowls**, **sand**, **water**, **oven mitts**, and a **temperature probe** connected to an **interface** that is connected to a **computer** with **Logger Pro**. Microwaves cook food by flipping water molecules so fast that the friction creates heat. **Looking at the materials and lab we will be using, what are the safety precautions we should take to protect ourselves and materials during this investigation?**

1) Microwaves cook food by flipping water molecules so fast (about 2.45 billion times per second) that the friction creates heat. Pour dry sand into a microwave bowl and check the temperature of the sand with the temperature probe. Write this data in Data Table 1.
2) Place the dry sand into the microwave and run the microwave for one minute. Using oven mitts, take the bowl of sand out of the microwave and measure the temperature with the probe. Write this data in Data Table 1.
3) Subtract the two numbers and write down the change in temperature in Data Table 1.
4) Place some more sand into another bowl and add some water. Measure the temperature of the sand and water before putting it in the microwave. Write this data in Data Table 1.
5) Put the bowl in the microwave and run it for one minute. Using your oven mitts, take the bowl of sand out of the microwave and measure the temperature with the digital thermometer. Write this data in Data Table 1.
6) Subtract the two numbers and write down the change in temperature in Data Table 1.

Data Table 1

	Temperature before Microwave (°C)	Temp after Microwave (°C)	Change in Temperature
Dry Sand			
Wet Sand			

Questions:

1) Which bowl had a greater temperature change?

2) Why do you think this happened?

3) How does this explain why dried food stuck to the inside of the microwave does not get hot?

4) Will a microwave always cook food? Explain.

Virtual Investigations that go with Waves

ExploreLearning.com:

    Waves Gizmo

    Ripple Tank Gizmo

    Phases Array Gizmo

    Sound Beats and Sine Waves Gizmo

    Longitudinal Waves Gizmo

    Hearing Frequency and Volume Gizmo

    Doppler Shift Gizmo

    Doppler Shift Advanced Gizmo

    Big Bang Theory – Hubble's Law

    Earthquakes 1 Recording Station Gizmo

    Earthquakes 2 Determination of Epicenter Gizmo

    Photoelectric Effect Gizmo

    Star Spectra Gizmo

    Herschel Experiment Gizmo

    Bohr Model: Introduction

    Bohr Model of Hydrogen Atom Gizmo

PhET.colorado.edu:

    Fourier: Making Waves

    Normal Modes

    Sound

    Wave Interference

    Wave on a String

Waves Intro

Band Structure

Bending Light

Blackbody Spectrum

Color Vision

Davisson-Germer: Electron Diffraction

Fourier: Making Waves

Geometric Optics

Lasers

Microwaves

Molecules and Light

Neon Lights and Other Discharge Lamps

Optical Quantum Control

Photoelectric Effect

Quantum Wave interference

Radiating Charge

Radio waves and Electromagnetic Fields

Simplified MRI

Physicsclassroom.com:

Physics Interactives:

Waves and Sound

Vibrating Mass on a String

Slinky Lab

Simple Wave Simulator

Wave Addition

Standing Wave Maker

Beat Patterns

Light and Color

Electromagnetic Spectrum Infographic

RGB Addition

Paint with CMY

Color Shadows

Filtering Away

Colored Filters

Stage Lighting

Viewed in Another Light

Young's Experiment

Concept Builders:

Waves and Sound

Wave Basics

Wavelength

Waves – Case Studies

Rocking the Boat

Wave Interference

Decibel Scale

Name That Harmonic: Strings

Name That Harmonic: Open-End Air Columns

Name That Harmonic: Closed-End Air Columns

Light and Color

Spectrum

Light Intensity

Color Addition and Subtraction

If This. Then That: Color Subtraction

Color Pigments

Color Filters

Middle School Physical Science NGSS Correlations

Properties of Matter Concept Map MS-PS1-3

Physical and Chemical Changes MS-PS1-23

Student Atomic Motion MS-PS1-4

Observing Molecular Motion MS-PS1-4

Seeing the Heating Curve (Physical Change) MS-PS1-4

Convection in Liquids and Gases MS-PS1-4

Which is Denser in the Mixture? MS-PS1-3

The Density of Oddly Shaped Objects MS-PS1-3

Virtual Investigations that go with Properties of Matter MS-PS1-34

Mixtures and Properties of Water Concept Maps MS-PS1-23

Elements Compounds and Mixtures Research MS-PS1-23

Elements Compounds and Mixtures MS-PS1-23

Separating Mixtures MS-PS1-23

Separating Pigments MS-PS1-23

50 + 50 Does Not Equal 100 MS-PS1-23

Percent Sugar in Bubble Gum MS-PS1-23

Making Ice Cream MS-PS1-23

The Conductivity of Electrolyte Mixtures MS-PS1-23, 3-4

Sugar or Salt MS-PS1-346, 2-3

The Solubility of Gas in a Liquid MS-PS1-34

Speed of Dissolving Solutes Lab MS-PS1-34

Heat and Saturating Solutions MS-PS1-34

Building a Model of a Water Molecule MS-PS1-1

Checking Polarity MS-PS2-3

Celery Transport MS-PS2-23

Transpiration Pull MS-PS2-23

Seeing a Stoma MS-PS2-23

How does Rain Form? MS-PS1-4

Virtual Investigations that go with Mixtures and Properties of Water MS-PS1-134, 2-23, 3-4

Structure of Matter Concept Map MS-PS1-1

Scale Model of a Hydrogen Atom MS-PS1-1

Model of an Atom Showing the Illusion MS-PS1-1

Building Bohr Models MS-PS1-1

Finding the Period in Periodic Table MS-PS1-12

Metal or Nonmetal MS-PS1-23

Periodic Table Activity MS-PS1-123

Making a Graphite Light Bulb (A) MS-PS1-234, 3-5

Making Models of Compounds MS-PS1-1

Making Molecular Models MS-PS1-1

Virtual Investigations that go with the Structure of Matter MS-PS1-123

Chemical Reactions Concept Maps MS-PS1-256

Temperature and Reaction Rates MS-PS1-6

Observing a Catalyst MS-PS1-23

Change in Temperature in Chemical Reactions MS-PS1-6

Removing Carbon from Sugar MS-PS1-256

Conservation of Mass in Equations MS-PS1-5

Home Chemistry MS-PS1-256

Types of Chemical Reactions MS-PS1-256

Conservation of Mass MS-PS1-5

The Law of Conservation of Mass MS-PS1-5

Conservation of Life: Photosynthesis and Respiration MS-PS1-5

Virtual Investigations that go with Chemical Reactions MS-PS1-2356

Acids and Bases Concept Map MS-PS1-2

Which is an Acid and Which is a Base? MS-PS1-23

A Homemade Indicator MS-PS1-23

Observing Acid Relief MS-PS1-23

Acid or Base Grape Juice Indicator MS-PS1-23

Characteristics of Acids and Bases MS-PS1-23

Which will Corrode a Nail? MS-PS1-23

Virtual Investigations that go with Acids and Bases MS-PS1-23

Motion Concept Map MS-PS2-2

The Motion of a Bowling Ball MS-PS2-2

Marbles in Motion MS-PS2-24

Ball Bounce MS-PS2-24

Cart on a Ramp MS-PS2-24

Picket Fence Free Fall MS-PS2-24

Elevator Lab MS-PS2-24

Virtual Investigations that go with Motion MS-PS2-2

Forces Concept Map MS-PS2-245

The Human Table MS-PS2-24

Balancing Forks MS-PS2-24

Comparing Friction Lab MS-PS2-24

Measuring the Effects of Air Resistance MS-PS2-24

Air Resistance MS-PS2-24

Observing Inertia, Newton's First Law of Motion MS-PS2-2

Inertia Lab Stations MS-PS2-2

Observing Centripetal Force MS-PS2-24

Centripetal Force Under Glass MS-PS2-24

Newton's Relay Race MS-PS2-12

Newton's Second Law MS-PS2-2

Fan Cart Lab MS-PS2-2

Newton's Third Law MS-PS2-1

Water Bottle Rockets MS-PS2-124

Virtual Investigations that go with Force MS-PS2-124

Mechanical Energy Concept Map MS-PS3-125

The Energy of a Pendulum Lab MS-PS3-125

Energy and Rockets Lab MS-PS3-12

Analyzing Elastic Potential Energy MS-PS3-125

Happy and Sad Balls MS-PS3-125

Conservation of Energy in a Toy MS-PS3-125

The energy of Colliding Objects MS-PS3-125

Who's got the Power? MS-PS2-24, 3-5

Levers Lab MS-PS2-24, 3-5

Simple Machines Lab MS-PS2-24, 3-5

Pulley Lab MS-PS2-24, 3-5

Bicycle Lab MS-PS3-1

Virtual Investigations that go with Mechanical Energy MS-PS3-125

Thermal Energy Concept Map MS-PS3-4

Energy Transformation Balls MS-PS3-45

Convection in Liquids and Gases MS-PS3-5

Observing Conduction Convection and Radiation MS-PS3-5

The Direction Thermal Energy Moves MS-PS3-45

Observing Molecular Motion MS-PS1-4, 3-45

Seeing the Heating Curve (Thermal Energy) MS-PS1-4, 3-45

Solar Oven MS-PS3-3

Testing the Rate of Heat Movement MS-PS3-345

Virtual Investigations that go with Thermal Energy MS-PS1-4, 3-345

Electromagnetism Concept Maps MS-PS2-3

Static Electricity MS-PS2-3

The Spinning Match MS-PS2-3

Charged Tape MS-PS2-3

Seeing Magnets MS-PS2-3

Making Electromagnets MS-PS2-3

Identifying Conductors and Insulators MS-PS2-3, 4-3

Battery Power MS-PS2-3

Making A Graphite Light Bulb (B) MS-PS2-3, 3-5

Human Circuits MS-PS2-3, 3-5

Comparing Series and Parallel Circuits MS-PS2-3, 3-5

Measuring Actual Voltage and Resistance MS-PS2-3, 3-5

Hotdog Circuits MS-PS2-3, 3-45

Motors and Generators MS-PS2-3, 4-3

Virtual Investigations that go with Electromagnetism MS-PS2-3, 3-45, 4-3

Waves Concept Map MS-PS4-12

Measuring Wave Properties MS-PS4-1

Observing Waves in a Slinky MS-PS4-12

Observing Sound MS-PS4-2

Coffee Can Phones MS-PS4-2

Music Test Tubes MS-PS4-2

Singing Glasses and the Dancing Toothpick MS-PS4-2

Playing the Rubber Band MS-PS4-2

Music has Patterns MS-PS4-123

The Doppler Effect MS-PS4-2

Making a Rainbow MS-PS4-12

Polarization of Light MS-PS4-2

3D Glasses MS-PS4-2

Light Pipes MS-PS4-2

Water Reflection MS-PS4-2

Test Tube Lenses MS-PS4-2

Reflection Lab MS-PS4-2

Magnifying Power MS-PS4-2

Brightness and Distance MS-PS4-2

Uses of the Electromagnetic Spectrum MS-PS4-3

How we use Microwave Ovens MS-PS4-3

Virtual Investigations that go with Waves MS-PS4-123

Equipment List for all Investigations

If you want to be able to do all the labs in this manual, here is the list of all the equipment you will need in order of appearance:

Water	Beakers
Scales	Vinegar
Meter sticks	Baking soda
Temperature probes	Steel wool
Interfaces	Beaker tongs
Computers	A clean piece of metal
Logger Pro	Corroded metal
Graduated cylinders	Refrigerator and freezer
Stopwatches	Hotplates
Rulers	Styrofoam cups
Pennies	Ring stands and clamps
Roll of pennies	Pepper
Empty penny rolls	Bottles
Safety goggles	Oil
Aluminum pans	Internet and textbooks
Paper	Aluminum foil
Pipettes	Milk
Candles	Laser pointer
Long neck lighters	Granite countertop samples
Matches	Kool-Aid
Salt	Pencil

Chalk	Nesquik
Muddy water	Gallon-sized Ziplock bags
Bar magnets	Quart-sized Ziplock bags
Horshoe magnets	Battery packs
Plastic baggies	Batteries
Wire strainer	Conductivity probes
Coffee filters	Christmas lights
Sand	Softsoap
Marbles	Orange juice
Iron filings	Sugar
Granola	Gatorade/Power-Aide
Scissors	Stirring rods
Pens and markers	Bottles of soda
Nail polish remover	Round balloons
Alcohol	Sugar cubes
Chromatography paper	Molecular model kit
Test tubes	Periodic Table
Test tube racks	Paper towels
Paper clips	Celery
Rubber stoppers	Food coloring
Bubble gum	White carnations
Gloves	Pressure sensor and tube set-up
Spoon	Plant branch
Crushed ice	Clear scotch tape

Lettuce	Hydrogen peroxide
Microscope slide	Small cups
Drinking glasses	Digital scales
Golf ball	Copper sulfate solution
Beads	Nails and screws
Film cases	Potassium iodide solution
Sulfur	Erlenmeyer flasks
Charcoal	Lead nitrate solution
Copper	Red cabbage
Mechanical pencil lead (graphite)	Shampoo
Colored pencils	Grapefruit juice
Glass jar with lid	Sprite
Wires with alligator clips	Ammonia
6-volt lantern batteries	100% purple grape juice
Blue tac	Antacid tablets
Alka-Seltzer tablets	Lemon juice
Aprons	Drain cleaner
Mentos tablets	Detergent
Epsom salt	Blue litmus paper
Borax	Red litmus paper
Laundry detergent	Universal indicator pH paper
Lighter fluid	pH meters
Ceramic bowl	Cokes
Liver	Bowling balls

- Large pillow
- Masking tape
- Flat board
- Motion detectors
- Hot Wheels track
- Small stickers
- Wire baskets
- Large bouncy balls
- Spring carts
- Vernier's Dynamic Systems
- Picket fence
- Photogates
- Mass sets with hooks
- Forks
- Toothpicks
- Ice cubes
- Rocks
- Erasers
- Wooden blocks
- Trays
- Toy cars that winds-up when you pull back
- Rubber bands
- Ping pong balls
- Ping pong paddles
- Index cards
- Tennis balls
- Buckets
- Wire coat hangers
- Brooms
- Basketball
- Carts and fans
- Dual-range force sensors
- Low g accelerometers
- Water bottle rocket launcher
- 2-liter bottles
- Air pump with pressure gauge
- String
- Stomp air rocket launcher
- 2lb & 8lb medicine balls
- Happy and sad balls
- Plastic shoe boxes & lids
- Toys using simple machines
- Baseballs
- Softballs
- PVC pipes
- Fulcrum collars & support stands
- Shoes
- Screwdrivers

Round and lever doorknobs

Pulleys

Gear changing bicycle

Steel energy transformation balls

Jiffy Pop popcorn

Hot air popper

Unpopped popcorn

Microwave popcorn

Microwave oven

Pizza boxes

Plastic wrap

Black plates

Food to cook

Tumblers

Pitcher

Electroscopes

Plastic combs

Nickles

BBs

Small plastic containers

Bolts

Current Conductor

Multimeter

Extension cords with alligator clips

Hotdogs

Pickles

Simple motors

Flashlight generators

Telephone cords

Slinkies

Coffee cans

Digital keyboard

Microphone probes

Whistling football

Water hose

Spray bottle

Worklight

Polarizing filters

Light sensors

3D glasses

Fiber optics

Square tanks

Flat mirrors

Protractors

Magnifying glasses

Oven mitts

www.ingramcontent.com/pod-product-compliance
Lightning Source LLC
Chambersburg PA
CBHW062348220526
45472CB00008B/1744